Carbon Capture and Storage in the United Kingdom

This book is a concise but comprehensive guide to the history, present and possible futures of carbon capture and storage policy and action in the United Kingdom (UK).

There have been multiple failed starts, promises and "last chances" for carbon capture and storage (CCS) in Europe, North America, China and Australia, but thus far it has repeatedly collided with the political and economic realities that the technology is too expensive and complicated to gain and keep policymakers' support. However, in the UK that might be changing, with explicit government support for CCS to help decarbonise industry. Set within the broader context of global interest in CCS, this book first outlines the technologies involved in the types of capture technology, transport options and storage options in the UK. It then briefly introduces an overarching policy analysis framework (John Kingdon's multiple streams approach) and uses it to give an account of the long history of CCS interest and efforts in three chapters covering the 1970s to 2002, 2003 to 2015 and 2016 to the present day. Marc Hudson focusses on the various arguments made for the introduction of CCS, and the slowly shifting coalitions of actors who make those arguments, while contrasting these with the perspectives of those opposed to CCS.

This book will be of great interest to students, scholars and policymakers researching and working in the field, as well as the related areas of energy policy, energy transitions and climate change.

Marc Hudson was educated in Australia, the United Kingdom and the United States. He worked as an aid worker in Southern Africa and as a physiotherapist in the UK, specialising in amputee rehabilitation, before undertaking a PhD at the University of Manchester. His thesis examined the strategies and tactics of incumbents resisting carbon pricing in Australia in the period 1989–2012, as a contribution to the study of the politics of socio-technical transitions. He has held postdoctoral roles at Keele and Sussex universities. His academic articles have appeared in *Environmental Politics, Energy Research & Social Science, Energy Policy* and other journals. He has also written for *The Conversation, Peace News* and *New Internationalist* and engaged in local climate activism for 15 years in Manchester. He believes that historical perspectives almost invariably deepen understanding of current events and has created and maintained a climate histories website called *All Our Yesterdays*, which can be found at allouryesterdays.info.

Routledge Focus on Energy Studies

Limits to Terrestrial Extraction
Edited by Robert E. Kirsch

Network Governance and Energy Transitions in European Cities
Timea Nochta

Residential Electricity Consumption in Urbanizing China
Time Use and Climate-Friendly Living
Pui Ting Wong and Yuan Xu

Carbon Capture and Storage in the United Kingdom
History, Policies and Politics
Marc Hudson

For more information about this series, please visit: www.routledge.com/
Routledge-Focus-on-Energy-Studies/book-series/RFENS

Carbon Capture and Storage in the United Kingdom

History, Policies and Politics

Marc Hudson

Routledge
Taylor & Francis Group

LONDON AND NEW YORK

First published 2024
by Routledge
4 Park Square, Milton Park, Abingdon, Oxon OX14 4RN

and by Routledge
605 Third Avenue, New York, NY 10158

Routledge is an imprint of the Taylor & Francis Group, an informa business

© 2024 Marc Hudson

British Library Cataloguing-in-Publication Data
A catalogue record for this book is available from the British Library

ISBN: 978-1-032-60911-9 (hbk)
ISBN: 978-1-032-60912-6 (pbk)
ISBN: 978-1-003-46106-7 (ebk)

DOI: 10.4324/9781003461067

Typeset in Times New Roman
by Apex CoVantage, LLC

Contents

Acknowledgements

I would like to express my thanks to the following people: Mark Sullivan, for endless relevant articles and links, chivvying and proofreading; Sam Gunsch, for encouragement and proofreading; David Pope, for kind permission to use his brilliant cartoon; Marc Roberts, for drawing me a bath; Jon Gibbins for permission to use the image from his 2007 PowerPoint presentation in Chapter 4; Matthew Lockwood, for permission to use an image in Chapter 5 he created; Nils Markusson, for an electronic copy of a useful resource; and all those people interviewed in 2022 for an IDRIC project on the politics of industrial decarbonisation. All errors and omissions remain my responsibility.

And finally, to Sarah, for everything.

Figures

Acronyms

BECCS	Bio-Energy Carbon Capture and Storage
BEIS	Business Energy and Industrial Strategy
BIS	Business Innovation and Skills
CCA	Climate Change Act
CCS	Carbon Capture and Storage
CFCs	chlorofluorocarbons
COP	Conference of the Parties
CCSA	Carbon Capture and Storage Association
CEGB	Central Electricity Generating Board
CEO	Chief Executive Officer
CfD	Contracts for Difference
CGS	Clean Growth Strategy
CRE	Coal Research Establishment
DAC	Direct Air Capture
DECC	Department of Energy and Climate Change
DEFRA	Department for the Environment, Food and Rural Affairs
DTI	Department of Trade and Industry
EMR	Energy Market Reform
EOR	Enhanced Oil Recovery
ETSU	Energy Technology Support Unit
EU	European Union
FEED	Front end engineering and design
FID	Final Investment Decision
HMT	His Majesty's Treasury
IDRIC	Industrial Decarbonisation Research and Innovation Centre
IASSA	International Institute for Advanced Studies Analysis
IEA	International Energy Agency
IEAGHG	International Energy Agency Greenhouse Gas Programme
IEEFA	Institute for Energy Economics and Financial Analysis
IGY	International Geophysical Year
IPCC	Intergovernmental Panel on Climate Change
IPIECA	International Petroleum Industry Environmental Conservation Association

MSA	Multiple Streams Approach
NIC	National Infrastructure Commission
OECD	Organisation of Economic Co-operation and Development
OCCS	Office for Carbon Capture and Storage
RCEP	Royal Commission on Environmental Pollution
SSE	Scottish and Southern Electricity
SNP	Scottish National Party
TUC	Trades Union Congress
UKCCSRC	United Kingdom Carbon Capture and Storage Research Centre
UNEP	United Nations Environment Programme
UNFCCC	United Nations Framework Convention on Climate Change
WMO	World Meteorological Organisation

1 Introduction: climate, technofixes and CCS

Contents

1.1 Aims of the book and intended readership

In August 2023 the British satirical and investigative magazine *Private Eye* described carbon capture and storage (CCS) as "the Loch Ness Monster of climate change mitigation: everyone can describe it; some believe in it; but most have yet to be persuaded it actually exists" (Private Eye, 2023). A few days earlier, the Scottish "Loch Ness Centre" had gained international coverage by calling on volunteers to come to the Loch and join a two-day search[1] (Chasan, 2023).

This book will not help you find the Loch Ness Monster that is CCS. However, it will help you decide whether – and why – you "believe in it" or not. The word "monster" comes from the French *monstre*, meaning to show. In the same way that Mary Shelley's Frankenstein shows us fears about the then-new technologies of electricity, and "meddling" in matters of life and death, a deeper knowledge of CCS and its tortured history can show us how human societies – especially the advanced capitalist ones of Europe and America – have tried to cope with the rising threat of climate change.

DOI: 10.4324/9781003461067-1

At the time of completing this book, UK business associations are clamouring for the government to make stronger, "louder" decisions in signalling support for CCS (Sheppard et al., 2023). In the words of a correspondent for *The Guardian*:

> Carbon capture has been hovering in the background for many years but only started to come to the fore as we reach the limits of what tinkering around the edges of business-as-usual can do to bring emissions down. Carbon capture is the preferred climate solution of the fossil fuel industry and its cheerleaders; those who stand to lose from the transformative changes needed to meaningfully tackle the crisis.
>
> (Church, 2023)

This book has two distinct but overlapping aims. First, it gives an account of the long and torturous history of carbon capture and storage *policy* development in the United Kingdom (UK) over the last several decades. That forms the basis of Chapters 3–5. Before that, this chapter outlines the climate problem, and the next introduces two analytic tools, the multiple streams approach (MSA) and hype cycles. Second, this book aims to give you some conceptual tools. I hope that these will help you when you are undertaking your own study and when you are exposed to "big claims" about proposals for enormous technological projects, such as CCS, hydrogen or geo-engineering.

This book is intended for anyone with an interest in CCS, from the "general reader" to those involved in more academic studies, as undergraduates or postgraduates.

Having explained what the book seeks to achieve, it is worth explaining what the book is *not* attempting to do, thus managing expectations and avoiding (or reducing) disappointment. There are three such disclaimers. First, it is not a detailed account of the *engineering* challenges for the various elements of CCS. There already exist many volumes that explore the challenges of capture, purification, transport and storage. See, for example, Bandyopadhyay (2014) and Baena-Moreno et al. (2023).

Second, this is not (I hope!) a partisan account of whether the British state has been either too slow or too enthusiastic in supporting the development of CCS. Different people will take different views on this (my views on CCS are explained near the end of this chapter), and the reader is invited to try to see the question from multiple perspectives.

Third, this does not purport to be a *global* history of CCS developments. While at various points I touch on developments elsewhere, a global history of the policy developments and setbacks around CCS is beyond the scope of this book.

1.2 Climate change

Climate change became a public policy *issue* in 1988 (the distinction between a problem and an issue will be explored in the next chapter). In the 35 years since then, the issue has come and gone from the headlines and the speeches of politicians and CEOs. Each wave of interest (discussed below) has seen television and radio programmes, books, protests and policy proposals. While concern is high, the level of public understanding of the exact mechanisms – and the urgency of action – is not as high as scientists and those who understand the science would like. I think that part of the problem is that the communication of the science is not well done. It often involves diagrams of the Sun and the Earth's surface, with labels of "infra-red" and "ultraviolet" radiation. I do not believe these images are particularly helpful, and below I set out two metaphors which will, I hope, help you understand (and then explain) not only climate change but also how CCS might fit into doing something about it.[2]

Those who want, for various reasons, to delay or indeed stop action on humanity's emissions of carbon dioxide (CO_2) never tire of saying, "*the climate is always changing*". This is a truism that often functions to divert discussion from what needs to be done and takes it down various rabbit holes and cul-de-sacs, thus achieving the delayers' goals. Climate is a famously contested term – for convenience it is defined in different ways, but it is different from the "weather in the desert is an average of the last 30 years" (Werndl, 2016). The people who make this empty statement fail to point out two things. First, the changes happening now are happening at a very high speed, in geological terms. Second, all of recorded human history (at least the last 12,000 years) has occurred during a relatively mild and stable period known to geologists as the Holocene.

Thanks to the various impacts of human activity (including, but also extending beyond, the role of carbon dioxide emissions), we have now left the Holocene and have entered the Anthropocene, the age in which human impacts are changing the planet's physical systems.[3]

This is beautifully captured in the following cartoon by award-winning Australian cartoonist David Pope, as seen in Figure 1.1 below.

Tolerably accurate, direct measurements of temperature exist from the middle of the 19th century (the first International Meteorological Congress, at which standardised measurements were agreed, was held in 1853). Before that, other proxies for temperature, such as tree rings, give an indication of how temperatures have changed. In 1999, a paper (Mann et al., 1999) demonstrating the rise in temperatures since the beginning of the 20th century was published, showing a "hockey stick" effect.

The usual baseline for both temperature and atmospheric concentration of carbon dioxide is the pre-industrial revolution, about 1750. It is estimated that since then the average global temperature has risen by 1.3 degrees.

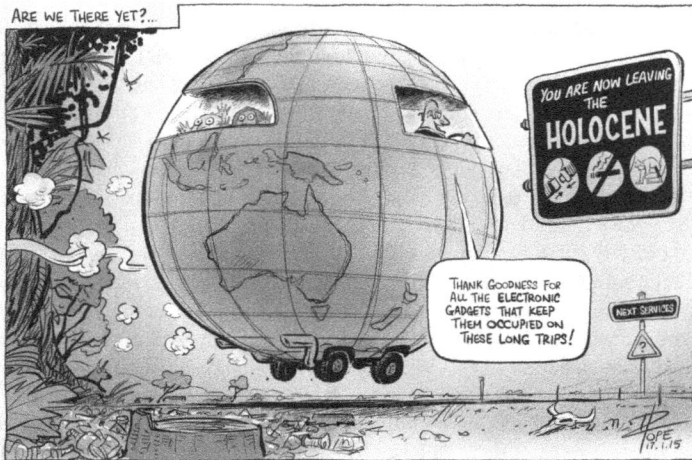

Figure 1.1 "You are now leaving the Holocene"

Global warming means that at a global level there will be warming, but it will be unevenly distributed. It does not mean that there cannot be localised episodes of cooling or that there will not be blizzards and snowstorms; that snow falls heavily in one place no more disproves climate change than someone eating a sandwich disproves a famine affecting millions of their compatriots.

The northern hemisphere summer of 2023, during which this book was written, has seen a number of climate records fall – around global average temperatures, Atlantic sea temperatures and Antarctic summer ice extent – to the alarm and disbelief of scientists and those who follow the scientific developments.

A note on terminology

In an attempt to counter the claim that "*the climate is always changing*", there was a vogue for referring to what emissions were doing to the planet's atmosphere as "Anthropogenic Global Warming" (AGW), where anthropogenic means "made by humans". In my opinion, the term is both clumsy and ugly, so I have chosen to refer to the changes in Earth's climate as climate change throughout the book. Other terms, such as "global weirding", "climate heating", "climate emergency", or "global boiling", have their merits but would be a distraction.

1.2.1 The two metaphors

There are many ways in which people attempt to explain climate change. In this section, I will introduce two metaphors, neither of which I invented. They involve different rooms in a house. Neither is perfect, with both requiring disclaimers. However, both have something to offer us in terms of mental maps for climate change and also for understanding the potential contribution of carbon capture and storage. The writing of these is deliberately in the tone of a "bedtime story".

1.2.1.1 In the bedroom

Imagine that you are lying in bed on a Sunday morning, under a thick blanket. You are not too warm, you are not too cold. In fact, you are, like the bear's porridge that Goldilocks feels titled to taste, "just right" (for Earth as a planet in the "Goldilocks Zone", see Overbye (2015)). You are feeling pretty comfortable in your own private Holocene. Now, if someone were to come into the room and pull off the blanket, saying, *"Get up, you lazy so-and-so"*, you would quickly feel too cold. You would start shivering in an effort to maintain a decent temperature.

Alternatively, if someone came into the room and said, *"you look so snug, let's give you another blanket"*, then for a very brief time you might enjoy the extra warmth, but pretty quickly you would be too hot and, in fact, dangerously overheating. You would start to sweat, which is your body's way of trying to lower the temperature by evaporation.

In this analogy you are the Earth's surface, while the blanket is the blanket of gases – especially carbon dioxide – that trap a certain amount of warmth from the sun. If the carbon dioxide level were to suddenly reduce, less heat would be trapped and the Earth's surface temperature would plummet.[4]

The analogy, like almost all analogies, is not perfect. There are two principal problems with it. The first is that the Earth does not generate significant quantities of its own heat, *pace* the geothermal engineers. In the metaphor, most of the heat you feel is generated by your own body, digesting the previous night's dinner, maintaining your core body temperature at around 37°C.

The second is that it imagines the doubling of the number of blankets as an instantaneous event. While 200 years (the time it will take humans to double the amount of CO_2 in the atmosphere) is indeed instantaneous from a *geological* perspective, it is also seven or so generations of humans. This is longer than most cultures (certainly Western cultures) are wont to consider. This second problem is dealt with in the next metaphor.

The reason that the Earth is overheating is that we are adding to that second blanket because every time we burn fossil fuels, namely oil, gas and coal,

Carbon dioxide concentration at Mauna Loa Observatory*

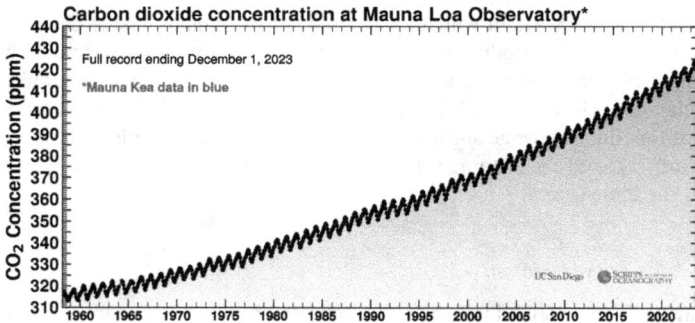

Figure 1.2 The Keeling Curve – Scripps Institution of Oceanography at UC San Diego

we release carbon dioxide as a byproduct of the combustion. Some of that carbon dioxide is removed from the atmosphere by natural processes into the oceans (which are both heating and becoming more acidic) or absorbed by trees and plants during photosynthesis.

These processes are no longer in absolute balance, thanks to the massive release of carbon dioxide in the form of fossil fuels that were laid down millions of years ago. Therefore, we are seeing a rise in atmospheric concentrations of carbon dioxide and other greenhouse gases, especially methane, which is much more effective at trapping heat, albeit for fewer years.

This build-up of carbon dioxide in the atmosphere, as we shall come to it, is measured by the famous Keeling Curve, see figure 1.2 above, named for the scientist Charles David Keeling, who began to take accurate measures of the atmosphere's carbon dioxide content in the late 1950s.

The upward curve reveals that – in the metaphor – a second blanket is being added.

The reason it is not a straight line is that in the Northern Hemisphere spring and summer large amounts of carbon dioxide are sucked out of the atmosphere as trees and plants grow. This is released in autumn and winter. Given that there is more vegetation in the Northern Hemisphere than the Southern Hemisphere, the atmospheric concentration oscillates within a year but remorselessly climbs across years.

What CCS aims to do is prevent a second blanket from being thrown onto you while you are lying in bed, or at least make the second blanket significantly less thick than the first. CCS would do this by reducing the amount of carbon dioxide vented into the atmosphere when fossil fuels are used, for either electricity generation or, as we shall see, especially in Chapter 5, industrial purposes.

1.2.1.2   The blanket metaphor

The blanket metaphor has two problems: the source of heat being the body, whereas it is the Sun "in real life", and the *speed* with which additional heat is trapped. For this, we move from the bedroom to the bathroom, and specifically the bathtub.

You cannot lie in bed forever, "bed-rotting" (Lee, 2023). At some point you have to get up, have breakfast, wash yourself. Instead of a quick – and environmentally more friendly – shower you choose to get in the bath. You put the plug in and open the taps. The bathtub begins to fill. You then get in the bathtub, and the level is enough to cover your knees. If you let the tap run indefinitely, the water will spill over the top of the tub, drenching the floor and ultimately leaking through the ceiling of the living room below. Your landlord will be most unimpressed. Therefore, you do not run the taps indefinitely. *Or* (and this is where the metaphor gets a little interesting), you could decide that you actually quite like the taps running fast. It makes you feel strong and modern, and you just like the sound of the water gushing and the constant new and fresh water surrounding you.[5]

The more systems-thinking among you may suggest that you can have your taps-open cake and eat it by pulling the plug out. Then, if you get it right, the amount of water from the taps will match that going down the plughole, and the level will not rise, and there will be no risk of flooding or an eviction notice from the landlord.

However, if things are not in balance – because either the plughole or the pipe below it gets blocked or the taps are turned on too full – then the water *will* build up in the tub. You can try drinking some of it, but that isn't going to help, is it? Trouble lies ahead.

Stepping back, for very large periods of the history of life on Earth, the level of water in the bathtub – or carbon dioxide in the atmosphere – has varied between 140 and 280 parts per million. It has oscillated within that range due to a number of factors, beyond the scope of both this metaphor and indeed this book. Humans have done two things that have affected the bathtub. First, if you remember, trees and plants draw down carbon dioxide – meaning the plughole functions to get rid of excess or "new" carbon dioxide. Over the last 2,000 years – and especially in the last 500 years – there has been massive deforestation, with forests cut down for fuel and building materials.[6] That means less carbon dioxide is being sucked out of the atmosphere.

Second, the oceans have been absorbing more carbon dioxide (and so becoming more acidic),[7] but as they become more acidic/less alkaline, their ability to soak up carbon dioxide is decreasing, rather like a sponge that is waterlogged is unable to soak up as much water as a dry one. The changing pH is also, of course, very bad news indeed for coral reefs, which also have warmer water, sea-level rise and, often, pollution to contend with.

So, the loss of trees and the changing ocean chemistry are as if you were sticking your little toe in the plughole, reducing the amount of water disappearing and causing the water level to rise slightly towards the brim of the tub.

Meanwhile, the amount of water (carbon dioxide and other greenhouse gases) entering the bathtub via the taps has shot up at an astonishing rate. The tap is turned ever more open, thanks to the ever-increasing amounts of oil, coal and gas burned for heat, transport and industrial purposes. Methane from rice paddies and cows adds to the mix.

These two factors – a partially blocked plughole and an increased amount of water flowing into the bathtub/atmosphere – mean that the water level is rising very quickly indeed. The danger of water overflowing the tub and causing mayhem is very high indeed as per Figure 1.3.

CCS in this metaphor can be seen as an effort to *decrease the amount of water flowing from the tap* (i.e. preventing it from getting into the bathtub at all – perhaps think of a jerry-rigged hose from the mouth of a tap bypassing the plughole and connecting directly to the outlet pipe). The more ambitious proponents of CCS, who advocate direct air capture (DAC; discussed in Chapters 5 and 6), can be thought of as suggesting having a bucket and throwing the water that is building up in the bathtub across into the bathroom sink, and so either slowing the increase of water in the tub or actually reducing it.

When thinking about CCS and the contribution it might make, think of this metaphor and the associated cartoon.

Figure 1.3 The bathtub and those who seek to keep it from overflowing

Renewables and the bathtub

The use of low- or zero-carbon energy sources – wind, solar, geothermal, tidal – can be seen as a way of reducing the amount of water coming into the bathtub. For all the noise and newspaper coverage of the rapid growth in renewable energy (especially wind and solar), most energy and electricity are generated from the burning of fossil fuels, and this is "unmitigated. That is to say, renewable energy has so far been *additional* to fossil fuel use, rather than a replacement for it. That will probably change, but it is a matter of time, and time is short.

As with the blanket metaphor, there are problems. The bathtub metaphor suggests that everything is more or less fine until the bathtub overflows. This is, as should be clear to anyone reading the news, not the case. Without belabouring the point, if you are splashing about in the tub, playing with your rubber ducks (Boxall, 2009) and so forth, inevitably a certain amount of water is ending up on the bathroom tiles. The damage is being felt in many parts of the world, though in the West, happy in the bathtub, it has been easy to ignore this.

In the time it has taken you to read these two metaphors, let alone this book, millions of tonnes of carbon dioxide will be added to the atmosphere (the bathtub) of this planet, the only one in the universe that we can be sure is habitable. Almost everything we do in our daily lives, whether it is eating, driving, flying, is saturated with fossil fuel use, either directly or indirectly.[8]

At present, human emissions of carbon dioxide are around 54 billion tonnes a year (Forster, 2023). It is very likely that in the time it takes you to read this book new announcements around carbon capture and storage projects will have been made by companies or governments, ostensibly helping to tackle the "54 billion tonnes a year" problem. Some recent announcements are mentioned in the final chapters, but given the speed of developments, a book that tried to keep ahead of the latest developments would be out of date by the time you took it home from the bookshop or downloaded it from a website.

1.3   Climate change

This book is not about the science of climate change – the metaphors above will have to suffice. Nor is it about the history of climate science and policy. Nonetheless, the *outlines* of the history will help you explain or understand the context in which CCS has (eventually) come to be promulgated as a viable – or at least attractive – option. This is especially the case with Chapter 3, which covers the period from the early 1970s through to 2002–2003.

Questions about what influences the weather have, understandably, fascinated thinkers for millennia. The entire history of how weather (and climate) has been thought of lies outside the scope of this book. For our purposes, the story begins in 1824. In that year the French scientist Joseph Fourier argued that *something* must be trapping heat from the Sun because otherwise, given our distance from it, we could expect the planet to be much colder. Fourier did not explore which gases were performing this task of a glass house effect (Fleming, 1999). Over three decades later, in the 1850s and 1860s, carbon dioxide – or carbonic acid, as it is known when in solution – was named a greenhouse gas by a female American scientist, Eunice Foote, to little or no interest.[9]

The next well-known advance came from Svante Arrhenius, the Swedish scientist who would later win the Nobel Prize for Chemistry. In an 1895 lecture, Arrhenius argued that, over the course of several hundred years, carbon dioxide would build up in the atmosphere, thanks to man's industrialisation, that is, the burning of fossil fuels. This, Arrhenius said, would indeed warm the planet but was no cause for concern. Arrhenius assumed this process would take hundreds of years and produce a warmer climate in northern latitudes, meaning food could be grown on land currently buried in snow for large portions of the year.

Arrhenius' predictions were challenged by other scientists on the basis of incomplete experiments on the way that carbon dioxide would trap heat. His theory did not fall into complete disrepute but was largely ignored by most scientists (though not so in the popular press). In the 1930s the carbon dioxide theory was taken up by British steam engineer Guy Callendar, who worked on the question of global warming (based on temperature records from around the world) in his spare time. Callendar received a polite but largely dismissive reception when he presented his work at the Royal Meteorological Society in London in 1938. Until his death in 1964, Callendar continued to argue his case that the planet was warmer and that this was due to carbon dioxide build-up (see, e.g., Callendar, 1961).

After the war, movement towards acceptance of carbon dioxide theory started in May 1953 when Gilbert Plass, a Canadian physicist, told the American Geophysical Union that "*the large increase in industrial activity during the present century is discharging so much carbon dioxide into the atmosphere that the average temperature is rising at the rate of 1.5 degrees per century*" (see Kaempffert, 1953; Hudson, 2023).

By then, initial planning for an enormous international collaboration, the International Geophysical Year (IGY), was underway. The IGY was a study of the entire planet – its oceans, landmasses, atmosphere and cryosphere. Even before the results were in, popularisers of science were warning of the long-term impacts of carbon dioxide build-up. One example will suffice – in 1957 a book titled *Once around the Sun* contained the following warning:

> The greenhouse effect . . . could melt the polar ice-caps, so that London, Paris, and New York would all be inundated with salt sea-water. Scientists are already thinking seriously of ways of keeping track of this man-made phenomenon, so as to be able to predict what the future has in store.
>
> (Fraser, 1957: 37)

After the IGY public attention was limited, but through the 1960s, measurements of carbon dioxide build-up continued, and computer models of not just weather but also climate began to improve (Edwards, 2013). In 1971, an international group of scientists met for three weeks in Sweden for the study of man's impact on climate, ahead of the 1972 United Nations Conference on the Environment in Stockholm. While this conference was relatively poorly attended (boycotted by the Soviet Union and regarded with suspicion and indifference by developing countries), it nonetheless created the United Nations Environment Programme (UNEP). UNEP combined with the World Meteorological Organisation (WMO) to put carbon dioxide build-up as a major item of investigation.

Throughout the 1970s, it became clear that the Earth was not, despite what a small number of scientists and journalists were saying, heading for a new ice age[10] but rather the opposite. In 1975, scientists met in Norwich, England, and concluded that an ice age was not coming but, rather, a warming. There were various largely unsuccessful attempts in the late 1970s to raise the alarm (covered in slightly more detail in Chapter 3).

In 1985, there was a meeting of scientists, organised by the WMO and UNEP, in Villach, Austria. This meeting was pivotal primarily because it challenged the assumption that carbon dioxide was the only significant greenhouse gas. The scientists realised that the long-anticipated warming would come faster than had been thought (Pearce, 2005). Scientists began to try to seriously interest policymakers in the issue. They met with most success in the United States and Australia. In December 1985, the famous American astrophysicist and science communicator Carl Sagan briefed US senators. There was a flurry of workshops and meetings among scientists and interested politicians. In 1988, the issue of climate finally burst onto the public policy agenda, thanks to several factors, including a prolonged drought in the United States, well-publicised testimony from NASA scientist James Hansen, an international conference on the Changing Atmosphere in Toronto and finally speeches by prominent politicians such as George H. W. Bush and British Prime Minister Margaret Thatcher.[11]

After resistance from the United States about the very idea of a treaty, negotiations began in early 1991 in Virginia (Matthews, 1991). They continued over five meetings held over the following year (Paterson, 1996). The United States and its allies resisted a treaty text which included targets and timetables for wealthy countries to reduce their emissions, as proposed by France. The US president, George H. W. Bush, stated that he would not attend the June 1992 "Earth Summit" if it included targets and timetables. Following the diplomatic intervention of the UK, the final text, agreed at a special last-ditch meeting in New York, did not include targets and timetables but merely an aspiration that nations would seek to stabilise their emissions at 1990 levels by the year 2000.

Article Two of the United Nations Framework Convention on Climate Change (UNFCCC) calls on the signatories to take action to ensure

> stabilisation of greenhouse gas concentrations in the atmosphere at a level that would prevent dangerous anthropogenic interference with the climate system. *Such a level should be achieved within a time frame sufficient to allow ecosystems to adapt naturally to climate change, to ensure that food production is not threatened and to enable economic development to proceed in a sustainable manner.*
>
> (UNFCCC, 1992; emphasis added)

If the changes were taking place over millennia rather than decades, there might be time for ecosystems to adapt. But humans would probably either not notice or not take any action, and you would not be reading this book.

The Framework Convention, perhaps because it was relatively weak and imposed no specific obligations on any nation, was ratified by enough nations and became international law relatively quickly. The first "Conference of the Parties" (COP) took place in March and April 1995 in Germany. The main outcome was the "Berlin Mandate", an agreement that rich nations would finalise negotiations about beginning to reduce their emissions by the third COP. The third COP took place in Kyoto, Japan. A fractious meeting (see Leggett, 2001) produced the Kyoto Protocol, which called for industrialised countries to reduce their emissions by variable (relatively small) amounts during the period 2008–2012. Crucially, this would involve both international emissions trading and so-called joint implementation, by which rich countries could count emissions reductions they had funded in other countries as a part of their own national efforts.

In 2001, the new administration of George W. Bush announced that it was withdrawing from the Kyoto Protocol. Australia's Prime Minister John Howard followed a year later, despite Australia having received a "reduction" target of 108% and another concession that in effect made the target 130%. This withdrawal from Kyoto is, as I argue at the end of Chapter 3, a crucial impetus for increased interest in CCS.

This period is vital for understanding the history of CCS and the attention given to it. Some governments that were most noisy in their support for technological solutions – especially CCS and hydrogen – were trying to avoid domestic cuts and also slow the progress of international negotiations.

The finer details of COP history need not detain us. There are annual meetings, with the exception of 2020 because of the pandemic. Some of these matter more than others, as UK environment secretary Therese Coffey controversially stated (Abdul, 2022). There was a failed attempt in 2009 to extend/ replace the Kyoto Protocol. In November 2015, COP21 in Paris led to an agreement that nations would strive to take actions that would reduce emissions to a level that would hold the increase in the global average temperature to "well below 2°C above pre-industrial levels" and also pursue efforts to limit the temperature increase to 1.5°C above pre-industrial levels.

At COP26, held in 2021 in Glasgow, nations were supposed to arrive with strengthened targets if the promises they made in 2015 (their "nationally determined contributions") were not adequate to the task of reducing emissions.

To return to the bathtub metaphor – each nation is responsible for pouring some water into the bathtub. No nation is going to benefit from reducing its *own* flow if others do not also do the same. So no-one acts forcefully, and the level climbs.

1.3.1 Policy responses – taxes, trading schemes

Even the climate problem became an issue in 1988; there was intense scepticism among those studying carbon dioxide build-up about the ability to even delay – let alone halt – global warming. A 1983 report by the US Environmental Protection Agency titled "Can we delay a greenhouse warming?" gave an answer, in essence, of "probably not" (Seidel, 1983).

The view was that fossil fuels were so embedded as energy sources, so hard to replace with alternative sources, and the costs of switching so high, as were the incentives to behave as a "free rider" (deriving benefits from the actions of others while not taking action yourself), that co-ordinating effective responses across well over 150 nations – each guarding its sovereignty and with an eye to competitive advantage – would prove impossible.

Once the climate issue became prominent, there was initial optimism, at least in some quarters, that the problem could be addressed in the same way the ozone problem had been. Given that a reader would have to be at least 50 years of age to have a direct memory of the ozone problem, a recap is in order. Ozone is composed of three oxygen atoms weakly bound together. High up in the stratosphere, ozone acts as a filter, reducing the amount of ultraviolet radiation reaching the Earth's surface. There had been concerns in the late 1960s and early 1970s that the emissions from fleets of supersonic jets might deplete the ozone layer and lead to crop failures and skin cancers. While the supersonic jets did not appear in the numbers forecast, there were ongoing concerns and bans on aerosol cans that used chlorofluorocarbons (CFCs) which broke down ozone.[12] There, the matter rested until 1985, when scientists working for the British Antarctic Survey discovered that the ozone layer was now much thinner than previously measured. A flurry of activity then took place in various international organisations between governments. Fortunately, CFCs were manufactured by a very small number of companies, and there were viable substitutes. Agreements were made, including compensation payments for developing nations and differing deadlines for the phase out of CFC production. However, in contrast to CFCs, fossil fuels are far more deeply embedded in human systems, with far fewer viable alternatives. The hopes of a repeat of the ozone success were soon dashed.

The earliest sets of policy proposals revolved around imposing taxes. In the same way that alcohol and cigarettes are taxed in order to raise the price (or at least create grey markets and black markets!), the argument went, a price on carbon dioxide would send a signal through the market to investors. In order to maintain profits, companies would shift from high-carbon energy sources towards lower ones. In reality, this did not happen. This was partly because where carbon taxes were implemented, they were not at a high enough level to create a signal. Often, taxes and emissions schemes were delayed and then introduced with loopholes that dampened the signal. On the price signal to the market, see Robinson (1992: 223–224). Carbon markets remain a favoured tool of policymakers, and over the coming years there will probably be intense battles about how CCS – especially carbon dioxide removals from technologies such as DAC– can be integrated, with credits and certificates gained for carbon captured.

1.4   The coming of "technofixes" from the 1960s onwards

It is easy to forget now, with the experience of the last 60 years, but the period after the Second World War through to the early 1960s was one of – at least in government and industry circles – enormous optimism about the ability of science and technology to solve all problems. This period, during which production and consumption of a vast array of materials began to expand at a previously unheard-of rate, is now known by scholars as the Great Acceleration (Steffen et al., 2015). There would be – thanks to civilian nuclear reactors – electricity too cheap to metre. Epidemic diseases would be cured. Scientists, deploying computers and rationality, would solve previously intractable problems of food shortages, illness and much else. "Better living through technology" was the mantra. And so was born the language of "technological fixes", "technofixes" for short. Under the guidance of American nuclear scientists such as Alvin Weinberg, the attractions of technofixes were many. They offered policymakers, industry and other actors a way of sidestepping questions about social justice and other awkward questions.

This faith in technology was shaken by a series of incidents and accidents. Nuclear power was beset by cost overruns and well-publicised accidents.[13] Rachel Carson's pivotal contribution – the 1962 book *Silent Spring*, about the damage unrestrained use of pesticides and fertilisers could cause – cast a pall over the dreams of consequence-free technoscience (Carson, 1962). By the late 1960s the war in Vietnam brought questions about the uses of technology into sharper focus (the split between older scientists who tended to accept the rationale of various US administrations and younger ones who were opposed caused deep splits within institutional science in the United States) (Moore, 1999).

One useful way of thinking about this was developed by the sociologist Allan Schnaiberg. He made a distinction between *production* science (aimed at increasing production – e.g. nuclear power, genetic modification, nanotechnology) and *impact* science, which measures the effects of production science on the natural world (including humans) (Schnaiberg, 1980.)[14]

Crucially, by the early 1970s the hopes of those advocating technofixes were shown to be at best overly optimistic and at worst wildly wrong. Many problems proved much tougher to resolve than had been thought in the heady days of the 1950s.[15]

The history of technofixes for climate change will be discussed in more detail in Chapter 3 (see also Ruser & Machin, 2016). In essence, they date back to the late 1980s with proposals for solar radiation management via mirrors in space, seeding clouds with sulphur to increase the reflectivity of clouds (so less sunlight reaches the Earth in the first place and cannot therefore be trapped underneath) and so on. The related question of the "hype cycle" will be introduced in Chapter 2.

1.5 The different technological options for carbon capture

Carbon capture and storage "does what it says on the tin".[16] In the following sections it is briefly discussed (the technical details and technological specifics lie beyond the remit of this book).

1.5.1 Capture

The first element is the capture of carbon dioxide at the source where it would otherwise be emitted. Until recently, most of the activity on this tended to focus on big sources – power stations, large industries. However, with the coming of the "net zero" imperative (discussed in more detail in Chapters 4 and 5), there has been increased interest in capturing from smaller point sources and indeed from the air or sea water.

Capture from point sources such as power stations can be divided into pre-combustion, post-combustion and oxy.

Pre-combustion involves converting fuel into a mixture of hydrogen and carbon dioxide before (thus "pre") it is burnt. After the carbon dioxide is separated and stored, the remaining hydrogen mix can be used.

Post-combustion, as the name indicates, involves separating carbon dioxide from the flue gases (flue meaning chimney or outlet pipe) after the fuel has been burnt. This is done by using a chemical solvent. As with oxy, described below, the equipment to do this can be added to new plants or retrofitted onto existing ones.

Oxy, as the name implies, involves burning a fuel in the presence of almost pure oxygen. This produces carbon dioxide and steam, with the steam being released harmlessly (in global terms – there can be issues around local thermal pollution).

Crucially, pre-combustion methods require significant modifications to power plants or factories, which may not have enough space for the retrofitting of pipes and other equipment. The distinction between pre- and post-combustion becomes important in the story of the first (failed) CCS competition in the UK in the period 2007–2011, described in Chapter 4.

As more companies enter the fray, as either potential customers or potential suppliers of capture technology, there is increased interest in other forms of capture, using algae, cryogenics (Aneesh & Sam, 2023) and other innovations. There is also increasing interest in so-called Capture as a Service, where intermediaries offer to take care of the logistics of both capture and transport.

1.5.2   Transport

Once the carbon dioxide has been captured, by whatever mechanism, it must be transported to somewhere it can be safely stored. (The question of the purity of carbon dioxide, and the implications of that purity for pipelines and shipping, is one of the many details that engineers and advocates need to consider, but beyond the remit of this book. For a discussion of the implications around different purity, see Splash (2023).)

There are several possible means of transport, including pipelines (either existing or newly built – see Cameron (2023), shipping and rail (or even road transport, depending on where the collection sources of carbon are and how far away the storage sites are).

The UK experience of CCS so far has been a story of the proposed re-use of existing pipelines and the construction of a small number of pipelines. While shipping was not part of the UK story from 1972 to 2023, it may become so, given that some areas of high carbon production do not have the possibility of using pipelines to transport captured carbon dioxide (e.g. the South Wales Industrial Cluster and the petrochemical industry around Southampton).

1.5.3   Storage

For CCS to make a meaningful contribution to preventing water from splashing out of the bathtub, vast quantities of carbon dioxide will need to be stored. As will be discussed in Chapter 3, the earliest proposal was simply to liquefy the carbon dioxide and pump it under pressure into the deep ocean, where it would stay there and not be a problem for humans. For various reasons, both legal and practical, this has fallen out of favour.

As far as the UK is concerned, for the period under discussion, the major interest has been in the North Sea and the Irish Sea. The North Sea Transition Authority (known as the Oil and Gas Authority until March 2022) estimates that there is a storage capacity of 78 gigatonnes in the North Sea. This may become a source of revenue, with nations, without the UK's storage capacity, storing their captured carbon dioxide here and creating new sources of revenue for the UK.

1.6 The author's perspective and blindspots

One time- and energy-consuming element of reading a book or article is trying to understand the biases and perspectives of the author. What are they emphasising, what are they minimising – or leaving out altogether – from their account of the topic at hand? Beyond their biases, what are their blindspots?

To help the reader avoid some of that work, here is a brief account of my perspective.

I grew up in a world that was aware of the dangers ahead – by the time I was 8 or 9 I was reading books and watching television programmes about environmental problems written from the early 1970s. Many focussed on the threat of extinction of "charismatic megafauna" and deforestation. By the mid-1980s, living in Australia, I was becoming sensitised to ozone depletion, which was especially salient for fair-skinned people fearing skin cancer. Along with many Australians, I first became properly aware of the "greenhouse effect" in 1989. By 1991, I had concluded that, as a species, we would not respond to this long-term threat with the determination and speed required. I first became aware of CCS as a "technofix" for climate change in the mid-2000s. I was involved in the first Camp for Climate Action, held near Drax power station in 2006. Before the Camp I was one of the authors of a book titled *Time Up*, which contained an interview on the subject with a Tyndall Centre researcher (Hudson, 2022).

I have had intense scepticism about CCS on purely pragmatic grounds. I struggle to believe that so much infrastructure could be built and then maintained at the speed and scale required. Further I worry that if somehow it *were* to succeed, we in the West would use it to give ourselves permission to continue with radically unsustainable patterns of behaviour. I was – and remain – of the belief that CCS has no role in propping up the coal industry and that money and expertise should be spent instead on energy efficiency, renewables and demand reduction overall. However, in the course of a postdoctoral post between late 2021 and early 2023 I became more fully exposed to the arguments for CCS around (as will be discussed more fully in Chapters 4 and 5) *industrial* decarbonisation. We will continue to require glass, steel, cement and so forth. While further incremental technological innovations may reduce their carbon footprint, it may well be that the only way to get these to "zero" is CCS. The dilemma, as I see it, is that the technologies, infrastructures and business models required for capturing the emissions from these sectors will only get built with the expertise and acquiescence of companies with the existing skills in chemistry, pipelines and transport of hydrocarbons – namely the oil and gas industry. And the oil and gas industry is going to use the existence of CCS to push for further exploration of what it sells – oil and gas.

In early 2022, presenting a webinar on my post-doctoral work, I used the famous image (Figure 1.4) of what might be a duck or a hare to illustrate that what a person might see in CCS depended on what they expected, wanted or needed to see.

A critic might say *"no – while that might be true of a picture, but in reality, you never mistake a duck and a hare. Carbon capture is"* and then follow this with a confident pronouncement.

Most days, I see CCS as what its critics say it is – an expensive and failed technology that has succeeded in providing "cover" for an industry that ought to have been swiftly regulated out of existence and "replaced with something nicer" as quickly as possible. On other days, I think "what about the steel, glass and cement?"

Figure 1.4 Duck or hare

Source: https://en.wikipedia.org/wiki/Rabbit%E2%80%93duck_illusion

I have spoken to many advocates for CCS. The ones I met (and this is of course not a random selection but rather those who would make time to talk to an academic researcher) were sincere, straightforward and not – as best I could tell – Machiavellians interested in prolonging the oil and gas industry but rather people concerned that industrial decarbonisation happens at pace and that the de-industrialisation of Britain is arrested and even reversed.

I have tried not to let my views influence my selection of facts or creation of a too-smooth narrative. You, reader, may disagree that I have succeeded, but at least now you know where I have been coming from.

1.7   The format of the rest of the book

Chapter 2 introduces two more framing devices. These are not metaphors like blankets or bathtubs but rather academic concepts – the MSA and hype cycles – that will structure the three empirical chapters that follow. In Chapter 3, I examine the growth in concern about climate change and the possible technological solutions to it, globally but especially in the UK. It is the story of the dog that did not bark in the nighttime on three separate occasions. In Chapter 4, I look at the real growth in interest in CCS globally, but especially in the UK, with a proposal from BP and two separate government competitions. This is mostly the tale of "let's save coal", but by the end of the period there is growth in interest in industrial decarbonisation. Chapter 5 tells the story of the incredible resurrection of CCS in the UK, after it might well have simply died a death after the dramatic withdrawal of government support in November 2015. It's a tale of determined and co-ordinated action by a (growing) coalition of actors. It is also the tale – since 2019 – of increased site-specific activity around preparing for the actual roll-out of CCS. The final chapter draws the strands together, suggests ways to keep abreast of developments (since the field is developing so quickly that the book will not be able to predict which projects come to fruition and which do not) and closes with some thoughts on what a "successful" roll-out of CCS would require and what "success" even means.

Notes

1   As fans of the BBC TV series *Doctor Who* may remember, the Loch Ness Monster was in fact a cyborg, used by stranded aliens, the Zygons, to destroy oil rigs in the North Sea. Sadly, the question of how human anxieties around energy – security of supply, geopolitics, consequences – manifest themselves is beyond the scope of this book!

2   Scientists who are asked by journalists to explain the science are often reluctant to use simple metaphors. They fear – probably with good cause – that doing so would be to risk their reputations. That is unfortunate, since good metaphors, properly flagged, can act as scaffolding that will help people's understanding.

3 The term "Anthropocene" is challenged as a mystification, eliding the actions of industrialised states and corporations into "anthro", meaning "man" or "humankind" (Luke, 2020).

4 In my last *Doctor Who* reference, I would point to a 1967 story, *The Ice Warriors*, as containing a startlingly scientifically illiterate version of this.

5 Please do not try this at home, especially given that large parts of the UK are in drought conditions and the water crisis is as serious as the climate crisis (Spear-Cole, 2023).

6 Intriguing work suggests that the collapse of North American civilisation after the arrival of Spanish fleets at the end of the 15th century, which introduced infectious diseases to which the populations had no resistance to, meant that lands that had been cleared for cultivation became dense forests again, lowering atmospheric carbon dioxide levels and contributing to the mini-Ice Age in Europe (Koch et al., 2019).

7 Those with a chemistry background reading this book will have winced at that phrasing. Technically, the oceans are becoming "less alkaline" – the pH is dropping towards "neutral" of 7.

8 This is not to say that climate change is only the fault of the individual Western consumer. There are lively, if deadening, debates about whose "fault" climate change is – individuals or systems. This is well beyond the scope of this book. For a good introduction on the "Imperial Mode of Living", see Brand & Wissen, 2021).

9 For more on the rediscovery of Foote's work, see Jackson (2020). Foote's work may have been appropriated by Anglo-Irish scientist John Tyndall (see Bell, 2021).

10 For more on the myth of the 1970s Ice Age "consensus", see Peterson et al. (2008).

11 Thatcher had been warned about climate change in May of 1979 by the UK's Chief Scientific Advisor.

12 For an interesting account of how manufactures of ozone worked to create doubt in the public mind, see Oreskes and Conway (2010).

13 The opposition of the US coal industry to nuclear power is fascinating, but beyond (far!) the scope of this book.

14 The broader question of what the late German sociologist Ulrich Beck referred to as "reflexive modernisation" is worth exploring when thinking about CCS but is beyond the scope of this book.

15 It is in response to this awakening that concepts like "wicked problems" and "post-normal science" came to be created.

16 The UK paint production company Ronseal ran a highly successful advertising campaign using this phrase, which has now entered the language (BBC, 2013).

References

Abdul, G. 2022. Up to King Charles whether he wishes to attend Cop27, says Thérèse Coffey. *The Guardian*, October 28. www.theguardian.com/environment/2022/oct/28/king-charles-free-to-attend-cop27-climate-summit-says-therese-coffey-rishi-sunak

Aneesh, A. and Sam, A. 2023. A mini-review on cryogenic carbon capture technology by desublimation: Theoretical and modeling aspects. *Frontiers in Energy Research*, 11. https://doi.org/10.3389/fenrg.2023.1167099

Baena-Moreno, F., González-Arias, J., Ramírez-Reina, T. and Pastor-Pére, L. 2023. *Circular Economy Processes for CO_2 Capture and Utilization Strategies and Case Studies*. Sawston: Woodhead Publishing.

Bandyopadhyay, A. (ed). 2014. *Carbon Capture and Storage CO_2 Management Technologies*. London: Routledge.

BBC. 2013. *The Ronseal Phrase: It Does Exactly What It Says on the Tin*, January 8. www.bbc.co.uk/news/magazine-20945352

Bell, A. 2021. *Our Biggest Experiment: An Epic History of the Climate Crisis*. London: Bloomsbury.

Boxall, S. 2009. From rubber ducks to ocean gyres. *Nature*, 459, 1058–1059. https://doi.org/10.1038/4591058a

Brand, U. and Wissen, M. 2021. *The Imperial Mode of Living: Everyday Life and the Ecological Crisis of Capitalism*. London: Verso Books.

Callendar, G. S. 1961. Temperature fluctuations and trends over the earth. *Quarterly Journal of the Royal Meteorological Society*, 87(371), 1–12.

Cameron, G. 2023. Removing North Sea oil structures to cost £40bn. *The Times*, August 10, p. 35.

Carson, R. 1962. *Silent Spring.* New York: Houghton Mifflin.

Chasan, A. 2023. Loch Ness Centre wants "new generation of monster hunters" for biggest search in 50 years. *CBS News*, August 6. www.cbsnews.com/news/loch-ness-centre-monster-nessie-search-new-generation-hunters-scotland/

Church, M. 2023. Rishi Sunak came to Scotland offering more North Sea drilling and carbon capture. We reject both. *The Guardian*, August 2. www.theguardian.com/commentisfree/2023/aug/02/rishi-sunak-scotland-north-sea-drilling-carbon-capture-we-reject-both

Edwards, P. N. 2013. *A Vast Machine: Computer Models, Climate Data, and the Politics of Global Warming*. Cambridge, MA: MIT Press.

Fleming, J. R. 1999. Joseph Fourier, the 'greenhouse effect', and the quest for a universal theory of terrestrial temperatures. *Endeavour*, 23(2), 72–75 (Fraser, R. 1957. Once round the Sun).

Forster, P. 2023. Greenhouse gas emissions are at an all-time high and Earth is warming faster than ever – report. *The Conversation*, June 8. https://theconversation.com/greenhouse-gas-emissions-are-at-an-all-time-high-and-earth-is-warming-faster-than-ever-report-207234#:~:text=Greenhouse%20gas%20emissions%20are%20at%20an%20all%2Dtime%20high%2C%20with,0.2%2C%C2%B0C%20per%20decade.

Fraser, R. 1957. *Once round the sun: The story of the International Geophysical Year*. London: Hodder and Stoughton.

Hudson, M. 2022. *Carbon Capture and Storage: What I Think I Think Today*, December 8. https://marchudson.net/2022/12/08/carbon-capture-and-storage-what-i-think-i-think-today/

Hudson, M. 2023. Climate change first 'went viral' exactly 70 years ago. *The Conversation*, May 12. https://theconversation.com/climate-change-first-went-viral-exactly-70-years-ago-205508

Jackson, R. 2020. Eunice Foote, John Tyndall and a question of priority. *Notes and Records*, 74(1), 105–118.

Kaempffert, W. 1953. How industry may change earth. *New York Times*, May 24, p. 11.

Koch, A., Brierley, C., Maslin, M. and Lewis, S. 2019. Earth system impacts of the European arrival and Great Dying in the Americas after 1492. *Quaternary Science Reviews*, 207, 13–36.

Lee, B. 2023. Bed rotting': What is this new tiktok generation Z self-care trend. *Forbes*, July 9. www.forbes.com/sites/brucelee/2023/07/08/bed-rotting-what-is-this-new-tiktok-generation-z-self-care-trend/gett

Leggett, J. K. 2001. *The carbon war: global warming and the end of the oil era.* Psychology Press.

Luke, T. W. 2020. Tracing race, ethnicity, and civilization in the Anthropocene. *Environment and Planning D: Society and Space*, 38(1), 129–146.

Mann, M., Bradley, R. and Hughes, M. 1999. Northern hemisphere temperatures during the past millennium: Inferences, uncertainties, and limitations. *Geophysical Research Letters*, 26(6), 759–776.

Matthews, J. 1991. Chantilly crossroads. *Washington Post*, February 10. www.washingtonpost.com/archive/opinions/1991/02/10/chantilly-crossroads/90b14241-3042-4a89-83a3-58eb38a3bfce/

Moore, K. 1999. Political protest and institutional change: The anti-Vietnam War movement and American science. In *How Social Movements Matter*, edited by Marco, G., Doug, M. and Charles, T. Minnesota: University of Minnesota Press, pp. 97–118.

Oreskes, N. and Conway, E. 2010. *The Merchants of Doubt.* London: Bloomsbury.

Overbye, D. 2015. As ranks of goldilocks planets grow, astronomers consider what's next. *Australian Financial Review*, January 7. https://www.afr.com/politics/as-ranks-of-goldilocks-planets-grow-astronomers-consider-whats-next-20150107-12jwmy

Paterson, M. 1996. *Global Warming and Global Politics.* London: Routledge.

Pearce, F. 2005. Histories: The week the climate changed. *New Scientist*, October 12. www.newscientist.com/article/mg18825210-600-histories-the-week-the-climate-changed/

Peterson, T., Connolley, W. and Fleck, J. 2008. The Myth of the 1970s global cooling scientific consensus. *Bulletin of the American Meteorological Society*, 89(9), 1325–1338. https://doi.org/10.1175/2008BAMS2370.1

Private Eye. 2023. Energy special. *Private Eye No.* 1604, August 11–24, p. 8.

Robinson, M. 1992. *The Greening of British Party Politics.* Manchester: Manchester University Press.

Ruser, A. and Machin, A. 2016. Technology can save us, can't it? The emergence of the 'techno-fix narrative in climate politics. *Conference: Technology + Society = Future?* Podgorica. www.academia.edu/28675395/TECHNOLOGY_CAN_SAVE_US_CANT_IT_THE_EMERGENCE_OF_THE_TECHNO_FIX_NARRATIVE_IN_CLIMATE_POLITICS

Schnaiberg, A. 1980. *The environment: From surplus to scarcity.* New York, Oxford University Press.

Seidel, S. 1983. *Can We Delay a Greenhouse Warming? the Effectiveness and Feasibility of Options to Slow a Build-Up of Carbon Dioxide in the Atmosphere.* Washington, DC: Environmental Protection Agency.

Sheppard, D., Millard, R. and Parker, G. 2023. Industry calls on UK to accelerate carbon capture as new projects approved. *Financial Times*, July 31. https://www.ft.com/content/d2e46191-282b-4718-af1f-1cfedfbf7d23

Spear-Cole, R. 2023. Water crisis should be considered as critical as climate change, researchers say. *The Independent*, August 23. www.independent.co.uk/climate-change/news/water-sustainability-china-uk-government-united-utilities-b2397894.html

Splash. 2023. *Not all C02 is the Same*, August 8. https://splash247.com/not-all-co2-is-the-same/

Steffen, W., Broadgate, W., Deutsch, L., Gaffney, O. and Ludwig, C. 2015. The trajectory of the anthropocene: The great acceleration. *The Anthropocene Review*, 2(1), 81–98.

UNFCCC, 1992. *United Nations Framework Convention on Climate Change.* New York. https://unfccc.int/resource/docs/convkp/conveng.pdf

Werndl, C. 2016. On defining climate and climate change. *The British Journal for the Philosophy of Science*, 67(2), 337–64.

2 Multiple streams approach and hype cycles

Contents

2.1 Introduction

This book encourages the reader to take a historical perspective on CCS in the UK. To aid that, the first chapter provided a short history of climate change and technofixes. This chapter introduces two further thinking tools – the MSA and hype cycles. This is followed by a brief outline of the political structures and policymaking processes of the UK.

2.2 Multiple streams approach

The MSA is the brainchild of American social scientist John Kingdon (Kingdon, 1984). He developed it while studying health policy in the United States. It is a popular tool for social scientists studying all kinds of policymaking for several reasons. One is that it provides a change to previous more "incrementalist" approaches. Another is that it is relatively easy to use.

The MSA has six headline elements (see Figure 2.1). Within each of those, there are sub-elements, which will be discussed in turn.

DOI: 10.4324/9781003461067-2

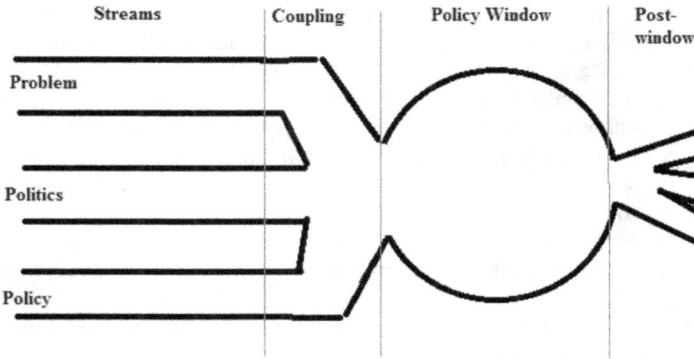

Figure 2.1 MSA

2.2.1 Streams

The MSA contains three streams of problem politics and policy. The "stream" metaphor suggests that momentum and turbulence might increase or decrease within a given stream, independently of the other streams.

2.2.1.1 Problem stream

The problem stream consists of four sub-components – load, indicators, focussing events and "feedback". Although the sub-components are not used systematically throughout Chapters 3–5, it is worth discussing each of these in a little more detail.

Load refers to the capacity institutions have to deal with problems. The presence of other (large) problems can make it harder for new problems to find an audience and to become "issues" (i.e. problems that are acknowledged by senior and influential figures as deserving time and energy towards their resolution). Jones et al. (2016: 15) note that "a new problem's ability to nudge its way into the purview of policymakers is negligible". Crucially, as we shall see in late 2015 (discussed at the end of Chapter 4), if other issues are more important for senior decision-makers, then a policy like CCS support can be almost casually dealt a near-fatal blow.

Indicators refer to the ability of actors to first identify and then monitor problems, including their severity. It has come to include unemployment rates, inflation or more specific metrics. For climate and energy policy, this might include the fluctuating mix of sources from which the UK derives its energy (and within that, its electricity).

Focussing events are events that make it easier for a problem to become an issue. They tend to be "jarring and sudden . . . attached to particular problems, providing powerful impetus for action or change" (Jones et al., 2016: 15). Focussing events might be scheduled or unexpected – an oil spill (as we shall see in Chapter 3), a war, a significant policy change in another country. They might include, for example, the annual COP meetings of the UNFCCC (especially if the UK is hosting the COP, as it did in 2021). Focussing events might also include scandals or physical disasters (none has so far happened with CCS).

Finally, feedback refers to reviews, reports or other "warning systems which monitor the problems and the effects of government policies" (Jones et al., 2016: 15).

Crucially, a problem does not become an issue until it has been turned into one by the determined and prolonged work of "problem brokers" (Knaggard, 2015). Problem brokers seek to turn something regarded as "normal" or "inevitable" into an issue that should be addressed. Typical actors within the problem stream include scientists, campaigners around a particular issue, crusading journalists and politicians looking to make a name for themselves by exposing a lack of sufficient action on a given issue.

2.2.1.2   The politics stream

The politics stream contains political parties and individual politicians. These politicians can be either in government or in opposition or behind the scenes (e.g. party strategists who are not themselves front-line politicians facing the electorate). These people are constantly looking for problems that are becoming – or could become – issues which they might use to their advantage. If in opposition, they might seek to either embarrass the government for not taking it seriously or set up their own stall in front of voters to say that they will deal with the issue differently and better.

The sub-components of this stream are political ideology and the "national mood". For the purposes of CCS, one important aspect of political ideology, which has shifted back and forth over the last 50 years, is the legitimacy of state intervention in the energy system (for whatever reasons) and, especially, in industrial policy (see below and also Chapters 4 and 5). Kingdon saw national mood as distinct from the results of opinion polls, which are focussed on the popularity of individual politicians and parties and perhaps a crude ranking of what issues seem to matter at a given time.

Beeson and Stone (2013: 3) explain that the politics stream is not a smooth flow but rather consists of "flows and ebbs" which reflect "changes in the national mood, the influence of public campaigns of interest groups, administrative or legislative turnovers and changes of allegiances of politicians in parliaments" (see below for discussion of flow within streams).

2.2.1.3 The policy stream

The final stream is the policy stream, in which policies which might provide solutions to a given issue are put forward.

The actors within this stream include trade associations (we shall see the crucial role of the Carbon Capture and Storage Association (CCSA), think tanks, university departments, pressure groups, trade unions, lobbying firms and so on). They all try – with lesser or greater resources and success – to put forward their solution or attack solutions that they don't like, which are gaining prominence and credibility.

Kingdon likened this stream to a "primeval soup", made up of policy ideas that have been propounded in the past and used or discarded. Kingdon drew on the "garbage can" model of Cohen et al. (1972). Cohen and colleagues made the obvious but unsettling point that policymakers – almost always short on time and information – do not start from "first principles" when confronted with a problem that has become an issue. Rather, they reach down into the metaphorical garbage can by the side of their desk, fish out crumpled-up balls of paper on which are written previous proposals that were abandoned, flatten them out and see if what was written there might work this time, given that it has been deployed to solve similar-seeming problems.

2.2.1.4 On the streams' longevity, their "flow" and the actors within them

The three streams will be fuller and emptier – according to circumstances – at any given time. It may be that they are never brought together and that a problem which might have solutions simply continues. Alternatively, some problems become issues and then are solved to enough people's satisfaction and become (or are perceived as) "non-problems". A good example is the ozone hole discussed in the last chapter.

There is a largely forgotten word – purl – that may be of use. It means to flow with a curling or rippling motion, as a shallow stream does over stone.

In the normal order of things (without a blockage that is then released or a large rock being thrown in the stream), there will be a relatively stable amount of flow and turbulence in a stream. This will depend on how many actors are trying to fill or empty the stream. (So, for example, those who deny human-caused climate change is happening, or say that it is a minor problem that can be adapted to, hope to reduce the volume and flow in the problem stream and, by extension, the politics and policy streams.)

Finally, there are some actors who are more comfortable and able to operate in one or another stream, or indeed only in one. Acting in any stream requires resources such as information, reputation and credibility and access to the media and decision-makers. To have all these usually requires significant amounts of money (but not always – as will be discussed with regard to Greenpeace's actions in 2008 in Chapter 4). Some actors tend to be more prominent in some streams than others, and this changes over time. For example, since 2014, charities and NGOs have been becoming extremely cautious about campaigning, when a new law required them to keep detailed records of their spending on campaigning in the year before an election. The Transparency of Lobbying Act, which passed by a single vote, has had a chilling effect on political communication by charities and pressure groups (O'Dowd, 2017).

In general, scientists have been reluctant to engage in the politics and policy streams. There are reputational risks for them in doing so, and most are keen not to be dismissed or become easily dismissable as "partisan". One interesting feature of the last few years is an increase in the number of natural scientists willing to speak out – and perform civil disobedience – around UK climate and energy policy (Capstick et al., 2022).

2.2.2   Coupling

Sometimes, the streams do come together, in a process known as coupling, performed by a "policy entrepreneur".[1] The policy entrepreneur must be someone with determination, credibility, resources and status who can show that the problem being spoken of in newspapers and on television and so forth is indeed real and pressing and that policy ideas found in the policy stream (or primeval soup) can be used by those in the politics stream to bring about some sort of resolution. At this point, a "policy window" (discussed below) opens.

In terms of climate change and the UK, the first successful policy entrepreneur would be Margaret Thatcher, who had been briefed repeatedly and tirelessly by diplomats like Crispin Tickell and scientists like John Houghton. In September 1988, three months after NASA scientist James Hansen had given testimony to the US Senate that the "greenhouse effect" had arrived, Thatcher gave a speech to the Royal Society about the issue. Coming from her, someone who had resisted strong UK action on acid rain, this was a surprise.[2] Another example of policy entrepreneurship will be covered in Chapter 5, when CCS returns from a near-death experience, thanks to the work of Minister of State Claire Perry.

2.2.3   Policy window

Once the streams have been coupled, a policy window opens. Windows tend to retain the attention of various political actors and media outlets for a relatively short period of time (opinions vary, but rarely do these last for more than a couple of years). Importantly, a policy window does not stay open until an issue is effectively dealt with. It may be that a policy window closes not because other issues have pushed it off the stage but because eventually it is seen as too difficult, and sheer exhaustion leads to enough powerful actors agreeing that it can or must be relegated to the "too hard" basket.

When a policy window does close after a policy has been agreed upon and formulated, a new period, of policy *implementation*, begins. There have been attempts to extend the MSA to implementation (see, e.g., Howlett et al., 2015), but these have been criticised as unhelpful (Breunig et al., 2016) and are not as intuitive as Kingdon's original framework.

The processes – of streams filling, of coupling and of battles in a policy window – can often leave residues. After a window closes, existing organisations may adopt the issue because it is close enough to what they already do, or else new organisations may spring up specifically devoted to this new issue. The story of CCS in the UK demonstrates this, with organisations entering the fray (e.g. the CCSA in 2006), while others, such as coal interests, depart (see Chapters 4 and 5).

Most of the time, the MSA is used to examine one particular episode of streams filling, coupling and then a window opening and closing. A longer approach can also be helpful. In my PhD thesis (Hudson, 2018), I used the MSA to trace the rise and fall of four public battles over carbon pricing in Australia between 1989 and 2011 (Figure 2.2).

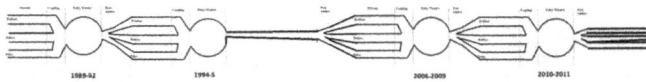

Figure 2.2 Sequential MSA streams and windows

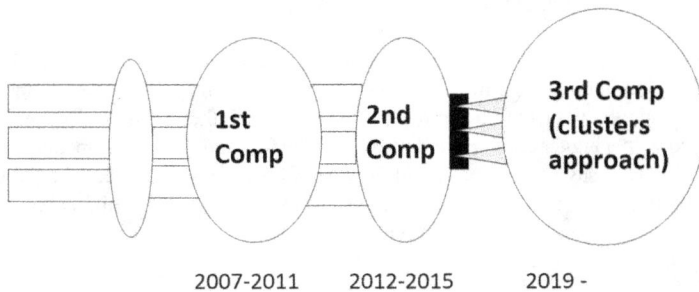

Figure 2.3 MSA and CCS policy in UK.

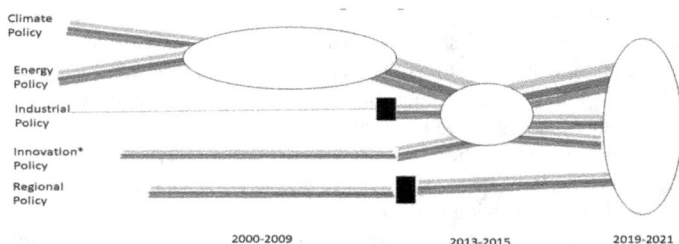

Climate
Policy

Energy
Policy

Industrial
Policy

Innovation*
Policy

Regional
Policy

2000-2009   2013-2015   2019-2021

Figure 2.4 Five UK policy domains beginning to combine and interact

There is one final use to which MSA - especially in this longitudinal sense – can be put. It can help show that policies are rarely completely isolated from other policies that fall under another heading.[3] As will be discussed in Chapter 4, climate policy and energy policy began to entangle between 2000 and 2009.[4] Meanwhile, industrial policy, long on the "outer" from the 1980s onwards began to stage a renaissance in the 2010s, in the aftermath of the Global Financial Crisis, as captured in Figure 2.4.

2.2.4   *Final provisos on MSA*

The MSA is best applied to issues that develop high political salience where the problem stream fills, thanks to the work of problem brokers, and a coupling follows.

It certainly applies to climate change *in general*. I would argue that there are three distinct policy windows in the UK climate history from 1988 to 1992, from 2006 to 2009 and then again from 2018 to 2021.

However, for CCS, which has rarely captured the public mind or attention, MSA is still useful, albeit slightly more unwieldy. The language of streams is fully used in Chapter 3. The language of windows is more loosely used in Chapters 4 and 5; as we will see, in Chapter 4, a kind of window forms in 2003, but it is with the coming together of the politics, the policy and the problem that an anaemic window gives rise to one project – Miller – which then folds, and in combination with a bigger climate window, we see the first and second competitions.

This is to say that the ebbs and flows of CCS are not in perfect sync with the climate issue more generally. This is unsurprising, given its highly technical nature, making it a topic on which most people have either little knowledge or a strong opinion.

2.3   Hype cycles

The hype cycle was developed by a Boston-based consultancy called Gartner. It has a similar intuitive usefulness to that of MSA.[5] There are many critiques of its applicability, and it certainly is not applicable to all technologies or as a predictive tool. For an application to CCS, see Arranz (2016).

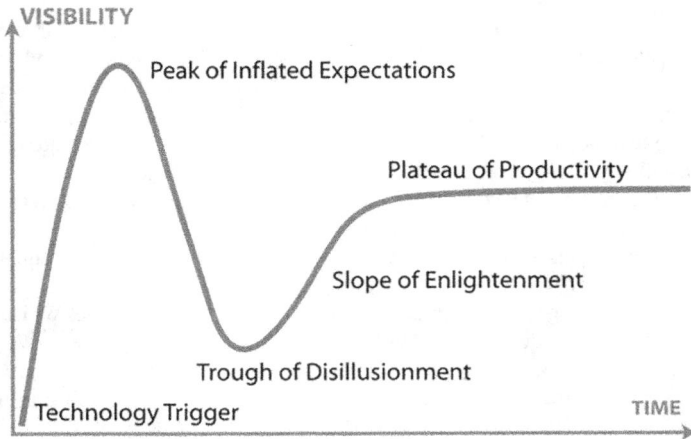

Figure 2.5 The hype cycle

It has, as per Figure 2.5, five key moments – technology trigger, peak of inflated expectations, trough of disillusion, the slope of enlightenment and plateau of productivity.

In the first phase, between the technology trigger and the peak of inflated expectations, there is a giddiness about what the technology will achieve. There are enthusiastic estimates of how quickly it will be ready for use, how quickly it will be adopted and how small and easily solved its teething problems might be. There is an incentive among proponents to make wild claims for its power and usefulness, the speed with which it can be rolled out and the speed with which it can outcompete existing technologies. This process feeds on itself because in order to get attention from a quickly jaded media, proponents make more and more extraordinary claims.

In this phase, reports about the future market share of the technology are produced and circulated on a seemingly weekly basis. They claim that via rigorous market analysis the market for [insert technology Zed] will be worth [insert enormous sum] by [insert date – usually the next three or four years]. These reports will be touted on websites and via press releases and will give rise to excitable and credulous short journalism.

More sceptical and battle-hardened (or world-weary, depending on your perspective) analysts find it harder to get a hearing in this period. These people, who have seen previous hype cycles, are at pains to point out the difficulties with the technology, the inevitable teething problems, the overblown claims, the underestimation of the effectiveness of the resistance of the incumbents who want to protect their assets and their revenue streams. Their warnings are often disregarded because "it's different this time",[6] and these analysts tend to be sidelined, ignored or mentioned as examples of people who are unduly pessimistic.[7]

2.3.1   *Plateau*

Once the deadline for the most outlandish claims has passed, or even approaches without sufficient signs of progress, then doubts creep in and previously ignored voices begin to get a hearing. There comes a point, usually within a year or two, when the technology is not progressing as fast as the more excitable boosters had predicted. Prototypes have more bugs and flaws than initially thought. Pilot schemes fail or reveal previously unthought-of difficulties and obstacles. The necessary investment money has not been gathered and final investment decisions (FIDs) begin to be deferred.

This brief period is followed by a kind of "anti-hype" period, an equally rapid period of renunciation. The difficulties of the technology that were glossed over or ignored or simply not known move to centre stage. Whistleblowers – disgruntled former CEOs or sacked employees – come forward, saying words to the effect that "I warned them that we couldn't go to market yet, but they wouldn't listen because they wanted to make a killing in the initial public offering". Remaining boosters of the technology try to dismiss these whistleblowers as malcontents with axes to grind. They protest that all technologies experience some setbacks, but the initial gloss and hope are now a distant memory.

At this point, a technology may simply vanish, to be dimly remembered as a trivia question or a punchline. Some technologies, however, will pass through this "trough of disillusionment" and – thanks to chastened proponents seeking redemption, or at least a return on their investment – come back in a much more specific, less economy-wide manner. Instead of being useful to "everyone all the time", a technology might find a niche or niches where it will be useful enough to some people for some purposes to be profitable. A good example of this would be the hype over Google Glasses. At one point, these were going to revolutionise our social lives and make learning and connecting ever easier. However, there was a backlash over concerns about privacy, and in any case, the devices themselves were found by some users to be clunky and cause headaches and nausea. This was not the end of the innovation, which has found niche rather than economy-wide applications (e.g. in healthcare training (Yoon et al., 2021).

As with the MSA, the hype cycle, while intuitive and seductive, should not be regarded as an iron law, a perfect template to drop onto a jumble of facts so that a pure story can be extracted. It is not the case that all technologies go through all stages.[8]

The particular relevance for CCS in the UK is to be found, especially in the period 2005–2008 and again from 2020 onwards. During these two upswings of hype, questions of scale and speed of roll-out have been too often batted aside, with the most optimistic estimates catching the eye.

Importantly, hype cycles also tell us that the rationale for a technology can change over time. This has certainly been the case with CCS. Initially (as we shall see in Chapters 3 and 4), it was framed as the salvation of coal (and perhaps gas) as sources of electricity generation in advanced industrial countries such as the United Kingdom, Australia and the United States. It did not happen, and coal is being pushed out of the electricity generation mix in those countries. In the UK, instead of coal, the rationale for CCS has shifted to industrial sources and carbon dioxide removal.

2.4 UK governance

Some readers of this book will be very familiar with the Westminster system of governance, and so they need not read the following closely.

The UK system of government, a constitutional monarchy also known as the "Westminster system", contains many features which will be familiar to people from other countries. However, this familiarity may offer false confidence about the UK system's particular quirks. What follows will only highlight aspects necessary for the explanation of the narrative in Chapters 3–5. There are many sources of information an interested reader can consult.

The House of Commons is made up of members of parliament (MPs) who are elected on a "first past the post system" in around 650 constituencies in the UK, which is made up of England, Wales, Scotland and Northern Ireland. An election must be called every five years, but it can happen more frequently, if the prime minister calls it or if the government "falls".

In most elections in living memory, one party won more than half the total number of constituency elections and was able to form a government without negotiating with any other party. The one exception is that between 2010 and 2015 there was a coalition of the Conservative Party and the Liberal Democrats, covered in Chapter 4.

The leader of the party with the most MPs in the House of Commons becomes the prime minister and has, as we shall see, enormous powers over the shape of government and the general direction of policy.

The central decision-making body of the government is the cabinet, appointed by the prime minister. This is made up of "Secretaries of State" with responsibility for different portfolios, in charge of a Civil Service department. Each Secretary of State will have several ministers of state (who can be from the House of Commons or the House of Lords – the unelected upper chamber). Some ministers may be invited to attend cabinet meetings – this is one of the powers the prime minister has to reward (or punish) ministers.

The prime minister can change his or her cabinet at any time (this is called a "re-shuffle"), though there have been cases where a Secretary of State has refused to move, calculating that the prime minister will decline to sack them.

The prime minister can also announce changes to the structure of the Civil Service and the Departments of State. Periodically, new departments are created (as the Department of the Environment was in 1970 and Energy in 1974). The departments may be abolished or combined – the Department of Energy and Climate Change in 2008. At various points in the CCS story, the rearranging of Departments of State and the appointment of Secretaries of State and ministers of state have mattered to the outcomes of policy.

Inevitably, there are jealousies, co-operation and power struggles of greater or lesser size between departments and Secretaries of State based on any number of factors. For the purposes of CCS, it is important to know that the Treasury (HMT) wields enormous power through its control over the overall budgets that each department has. The role of the Treasury, for better or worse, will become particularly clear in Chapter 5. However, at various times, other departments (Foreign Office, DECC and BIS, and later BEIS) have also been key players.

The prime minister can also create bodies such as royal commissions or, through legislation, independent advisory bodies such as the Climate Change Committee (see Chapters 4 and 5).

Governments introduce legislation, often based on so-called White Papers, which can be thought of as a relatively detailed discussion document that gives an indication of the government's future intentions.

The opposition parties seek to highlight government complacency, ignorance and incompetence in creating and implementing policy (this is where the MSA is particularly useful. Oppositions can choose which issues they seek to highlight and how to add to the streams.) Opposition parties look for problems that can become issues and try to create policy windows on topics that the government will struggle to act decisively. Opposition parties, via questions to ministers (and prime ministers' questions) and other mechanisms, try to build up a sense (part of the "national mood") that the country would be better positioned to deal with the challenges it faces if the government were replaced by the opposition party.

However, on some issues (including, until recently, climate change and, to a large extent, CCS policy), there has been a consensus between the major parties for quite some time (see Chapters 3, 4 and 5). In the second half of 2023, this consensus was undermined by Prime Minister Rishi Sunak, with potentially enormous consequences for CCS (discussed in Chapter 6).

Select committees are another crucial aspect of parliamentary business that is of importance for investigating and understanding the history of CCS in the UK. These committees, made up of MPs from all parties, largely mirror the work of Departments of State (i.e. when Departments of State are created or change, select committees do likewise). There are select committees in both the House of Commons and the House of Lords. The role of select committees is to scrutinise the work of the government in both policy formation ("do policies exist?") and the performance of both the government and the Civil Service in implementing existing policies. Select committees are not controlled by the prime minister or the government, and as shall be shown in Chapter 5, they play an important part in questioning government behaviour over CCS policy. Select committees can call witnesses from the government, business and civil society, and the reports they produce can have an effect on the direction of government action.[9]

Select committees are not the only means by which the work of the government and Civil Service is scrutinised. Newspapers and specialist media take a keen interest in what is happening. They obtain information via friendly sources, "leaks", Freedom of Information Act requests and other means. Alongside this, there are independent "watchdog" bodies, funded by the state but at least nominally independent of the government, which hold enquiries and produce reports. Two of particular relevance for the most recent history of CCS are the National Audit Office (NAO) and the Climate Change Committee (CCC), which will be discussed in more detail in Chapters 4 and 5.

The UK has historically been quite centralised. One interesting factor in the CCS story has been the role of regional actors. The first of these were the Regional Development Agencies, created in 1998 by the Blair government and abolished by the coalition in its "bonfire of the quangos" in 2010–2011. Since then, the devolved governments of Wales and Scotland and organisations such as the Teesside Collective (see Chapters 4 and 5) have played a crucial role.

2.5   Corporate political activity/public strategy

The role of business in attempting to influence governments and society more generally is vast and far beyond the scope of this short book. The reader is encouraged to read further about corporate "political activity". The reader is also encouraged to think of business not as a faceless monolith of utter unity and power but as a number of competing and overlapping groupings with shifting motivations.

Various sectors of the economy have trade associations (glass, cement, aviation, energy, retail, etc.). The roles of trade associations are various – to lobby government, to offer members information about the impact of proposed legislation and regulation, to provide networking opportunities and, occasionally, to launch publicity campaigns aimed at the general public about issues that affect members (this however is expensive, and members of trade associations can baulk at the costs).

In terms of lobbying the government, trade associations and individual companies (if they are large enough and are so inclined) will attempt to influence the government by supplying information and producing reports (often based on economic modelling) that call for government action to secure the conditions for profitability for sectors. This is usually couched in terms of "jobs created" or saved.

For the CCS story, the most crucial trade body has been the CCSA (created in 2006 – see Chapter 4). But this is not to say that many other national groups, such as the Confederation of British Industry, Make-UK and other bodies (such as the Aldersgate Group and Green Alliance), have not mattered.

2.6　Mousetraps, Machiavelli and teleology

In closing this chapter, there are three further points that may help the reader think about how technologies thrive or die.

First is the ease with which a new technology can be created and distributed. There is an expression that assures inventors that "*if you build a better mousetrap the world will beat a path to your door*". This is not necessarily the case. The most easily understood example of this is the Betamax video recorder, regarded as technologically superior to the rival VHS. Betamax lost out because the makers of VHS had secured a much wider array of films from Hollywood studios. While this is not a perfect analogy (energy supply is not like sitting down to watch a movie), it is worth remembering that the "superior" technology will not necessarily win – there are a host of factors around public acceptability and how well the technology "fits" with other (financial and cultural) demands and expectations.

Second, innovation is difficult. As the Italian political theorist Machiavelli put it in *The Prince*:

There is nothing more difficult and dangerous, or more doubtful of success, than an attempt to introduce a new order of things in any state. For the innovator has for enemies all those who derived advantages from the old order of things, whilst those who expect to be benefited by the new institutions will be but lukewarm defenders. This indifference arises in part from fear of their adversaries who were favoured by the existing laws, and partly from the incredulity of men who have no faith in anything new that is not the result of well-established experience. Hence it is that, whenever the opponents of the new order of things have the opportunity to attack it, they will do it with the zeal of partisans, whilst the others defend it but feebly, so that it is dangerous to rely upon the latter.

(Machiavelli, 1993)

Finally, more broadly, there is a tendency – related to technological fixes (as discussed in Chapter 1) – to believe that technology and history are headed in a specific direction and that because something "should" or "must" happen, then, as if by magic, it will. As we will see at the end of Chapter 4 and the beginning of Chapter 5, defeat can be snatched from the jaws of victory. There is an entirely plausible alternative history where CCS would no longer be part of the UK climate policy narrative.

2.7　Conclusion

The first two chapters have been "scene-setting", explaining the basics of climate change, of some of the tools that we can use to understand technological (and social) innovation.

In the following three chapters I outline the empirical story of CCS activity in the UK.

In the next chapter I trace the development of both UK government climate change awareness and the growth of interest in CCS from the late 1960s through to the year 2002, using Kingdon's streams. At this point, CCS emerges as a topic of considerable interest to policymakers. Chapter 4 traces three different efforts to get support for CCS development, including a proposal from BP and two different competitions. Chapter 5 deals with how the abrupt cancellation of the second CCS competition provoked a determined and coherent campaign to put CCS "back on the agenda", where it remains, for now. Chapter 6 discusses the implications of the history presented in Chapters 3– 5, possible futures for CCS in the UK and beyond.

Notes

1 The term "stream engineer" or "stream coupler" would be more helpful.
2 This demonstrates that the policy entrepreneur may not be the first person to talk about an issue – indeed, there would have been someone who was opposed to action but has since, but changed their stance, or been forced to do interest by circumstances and political calculations.
3 Thus, the constant pleas for "joined-up thinking" around, say, food policy and health policy, and the more general interest in "policy mixes" (Rogge & Reichardt, 2016).
4 Just because they converge, overlap and entangle does not mean that they are the same thing and do not retain independent characteristics. It is just that – at least until very recently – energy decisions were not made without at least lip service to climate concerns.
5 I would encourage readers to be cautious about such seductive and "intuitive" concepts.
6 The "16 Rules for Investment Success" (Templeton, 1993) is a classic. The following is by far the most quoted passage:

> "'This time it's different' is something of a sacrilegious phrase in the investment industry. Every time markets, corporate fundamentals or long-followed economic ratios enter above-average territory, there are sure to be pundits or investors warning about complacency and imminent mean reversion. Anyone who claims it's different this time is mocked with disdain for daring to question long-term financial relationships"
>
> (Carlson, 2017).

7 Perhaps surprisingly in an academic book, I would point the reader to an early episode of the US animated television programme *The Simpsons*, called Marge versus the Monorail. It shows the lonely battle Marge Simpson fights against hysteria in Springfield for an unnecessary and indeed dangerous Monorail!
8 Further criticisms can be found in Dedehayir and Steinert (2016).
9 For the purposes of flow, the nature and role of All Party Parliamentary Groups will be held back until Chapter 5.

References

Arranz, A. M. 2016. Hype among low-carbon technologies: Carbon capture and storage in comparison. *Global Environmental Change*, 41, 124–141.

Beeson, M. and Stone, D. 2013. The changing fortunes of a policy entrepreneur: The case of Ross Garnaut. *Australian Journal of Political Science*, 48(1), 1–14.

Breunig, C., Koski, C. and Workman, S. 2016. Knot policy theory. *Policy Studies Journal*, 44(S1), S123–S132.

Capstick, S., Thierry, A., Cox, E., Berglund, O., Westlake, S. and Steinberger, J. K. 2022. Civil disobedience by scientists helps press for urgent climate action. *Nature Climate Change*, 12(9), 773–774.

Carlson, B. 2017. This time really is different. *Bloomberg*, June 26. www.bloomberg.com/view/articles/2017-06-26/this-time-really-is-different?leadSource=uverify%20wall

Cohen, M. D., March, J. G. and Olsen, J. P. 1972. A garbage can model of organizational choice. *Administrative Science Quarterly*, 17(1), 1–25. https://doi.org/10.2307/2392088

Dedehayir, O. and Steinert, M. 2016. The hype cycle model: A review and future directions. *Technological Forecasting and Social Change*, 108, 28–41.

Howlett, M., McConnell, A. and Perl, A. 2015. Weaving the fabric of public policies: Comparing and integrating contemporary frameworks for the study of policy processes. *Journal of Comparative Policy Analysis: Research and Practice*. https://doi.org/10.1080/13876988.2015.1082261.

Hudson, M. 2018. *Enacted Inertia: Incumbent Resistance to Carbon Pricing in Australia 1989–2011*. Doctoral dissertation, PhD Thesis, University of Manchester.

Jones, M. D., Peterson, H. L., Pierce, J. J., Herweg, N., Bernal, A., Lamberta Raney, H. and Zahariadis, N. 2016. A river runs through it: A multiple streams meta-review. *Policy Studies Journal*, 44(1), 13–36.

Kingdon, J. 1984. *Agendas, Alternatives, and Public Policies*. Boston: Little, Brown.

Knaggard, A. 2015. The Multiple Streams Framework and the problem broker. *European Journal of Political Research*, 54(3), 450–465.

Machiavelli, N. 1993. *The prince (1513)*. Hertfordshire: Wordsworth Editions.

O'Dowd, A. 2017. Is lobbying legislation chilling public debate on health? *British Medical Journal*, 358. https://doi.org/10.1136/bmj.j4225

Rogge, K. S. and Reichardt, K. 2016. Policy mixes for sustainability transitions: An extended concept and framework for analysis. *Research Policy*, 45(8), 1620–1635.

Templeton, J. 1993. *Sixteen Rules for Investment Success*. www.franklintempleton.com/forms-literature/download/TL-R16

Yoon, H., Kim, S. K., Lee, Y. and Choi, J. 2021. Google glass-supported cooperative training for health professionals: A case study based on using remote desktop virtual support. *Journal of Multidisciplinary Healthcare*, 1451–1462.

3   From the 1970s to 2002

Contents

3.1   Introduction

This chapter covers the period from the very birth of British political concern about the potential problem of climate change to the first years of this century. The problem has become an issue, and promises of action have been made. Technological options around capturing carbon began to be explored (even enacted, in the case of the Norwegian Sleipner field). The chapter explores how CCS was cautiously explored (the word "championed" is too strong) by small groups of scientists and then adopted by those looking to continue to

DOI: 10.4324/9781003461067-3

use coal for energy generation in the UK. It is divided into four periods – from 1969 to 1979, during which there are some developments in CCS, but these are minimal, since the problem of climate change is not yet agreed upon as an issue. The second period, 1980–1989, shows concerns about climate change growing and finally breaking through among policy elites, with a rapid consideration of "what is to be done?" The third period, 1990–1997, sees the first recognisable flurry of scientific and technological interest in precisely what would be involved in capturing, transporting and storing carbon dioxide. Interestingly, "utilisation" is being talked about at this early stage. The final short period, 1998–2002, shows accelerating interest in CCS and the creation of various bodies that would prove to be important for its future.

3.2   Period 1: 1969 to 1979

When this period began, the atmospheric concentration of carbon dioxide was 324 ppm, with an estimated annual amount of carbon dioxide being emitted by humans at 13.77 billion tonnes. At the end of the period, atmospheric concentration was 337 ppm, with emissions at 19.62 billion tonnes per annum; to use the analogy of the bathtub from Chapter 1, the amount flowing into the bathtub was gradually increasing, as was the water level.

3.2.1   Problem stream

As discussed in Chapter 1, the potential dangers in a build-up of carbon dioxide in the atmosphere had been discussed by Arrhenius in 1896, by Callendar in 1938 and, with more consequence, by Plass in 1953 (Hudson, 2023). Through the 1950s and 1960s, scientists (including British scientists working in a variety of fields) became aware of the build-up and started factoring it into their work. By the late 1960s, with the rise of concerns about pesticides, water pollution, extinction of various species and population rise, the problem of air pollution moved from being a local issue, for example, smog, to the potential long-term problem of carbon dioxide build-up. This, from *The Listener*, in April 1969, is typical.

> There is another side to air pollution – a side verging so closely on science fiction that one would be tempted not to take it seriously, if it wasn't for the fact that the US government is building immensely complex computer models to predict its effects. Over the world as a whole 6,000 million tons of carbon dioxide – normally harmless – are poured into the atmosphere every year. In 30 years' time, there will probably be 25 per cent more carbon dioxide in the air than there is now. . . . One estimate is that this radical change in the atmosphere's heat balance could raise temperatures by between one and seven degrees Fahrenheit.
>
> (Fairley, 1969)

Later the same year, a writer in the *Financial Times* was able to wearily describe the issue as "the favourite cosmic disaster theory" (Winsbury, 1969). In August 1970 the potential problem of global warming was dismissed by the Alkali Inspectorate (the body charged with regulating, and reducing, pollution from industrial sources), achieving front-page news status in *The Times* (Wright, 1970)

Fortunately, the Alkali Inspectorate's dismissal of the issue did not prevent scientists from many countries continuing to work on the issue (Weart, 2008). Within a few years, an American oceanographer, in one of the first uses of the term "global warming", published a paper in the journal *Science* with the title "Climatic Change: Are We on the Brink of a Pronounced Global Warming?" (Broecker, 1975).

While American scientists were studying the topic (see, e.g., the pivotal 1977 study "Energy and Climate" (National Research Council, 1977), less was happening in the UK, for reasons discussed in Agar (2015) – see also Martin-Nielsen, 2018).

3.2.2 Politics stream

The earliest evidence of political concern about the possible problem of climate change comes from a mid-1968 memo by a minister in the Labour government. The memo, to the Secretary of State for Housing and Local Government, argued for the creation of some sort of monitoring body on environmental problems. Its author, Lord Kennet, included *"possible changes in macroclimate caused by the heating of the atmosphere due to industry"* in the list of items that might be monitored (Kennet, 1968 – see also Sims, 2016: 198).

The following year, around the world, environmental concerns moved from being a problem to an issue, and senior politicians responded. The UK Prime Minister, Labour's Harold Wilson, announced a standing Royal Commission on Environmental Pollution (RCEP) and, in May 1970, released the first Environment White Paper, which included a brief mention of the potential problem of carbon dioxide. Opposition leader Edward Heath criticised Wilson – not for indulging in a "green scare" but for not going far enough.[1] When Heath became prime minister after the June 1970 election, he created the first Department of the Environment. In its first report (RCEP, 1971), the RCEP mentioned the carbon dioxide issue as something that would require close monitoring. (Thirty years later the RCEP would play an important part in increasing emissions reductions ambition – discussed at the end of this chapter.)

One other incident here is worthy of mention. In July 1971, a Dutch ship, the *Stella Marris*, set sail from Rotterdam with the intention of dumping 600 tonnes of waste chemicals off the west coast of Ireland. Questions were asked in the House of Commons (Hansard, 1971), there were local protests and the ship returned to port with its cargo. This was the impetus for the London Convention for the Prevention of Marine Pollution by Dumping of Wastes at Sea. This question of what "waste" could be dumped where has importance for the CCS story, as we shall see in Chapter 4.

The next major event in the politics stream worthy of note was the 1973 Oil Shock. Prices for oil had been climbing through the first half of 1973 but quadrupled in a matter of weeks at the end of the year, when Arab oil exporters announced a boycott in retaliation for Western support for Israel in the Yom Kippur war. This event forced Western governments to re-examine assumptions of relatively cheap and abundant energy. It saw a flurry of new bodies formed. For the purposes of the CCS story, two are of particular note – the International Energy Agency (IEA) as a sister organisation to the Organisation of Economic Co-operation and Development (the OECD) and, in the UK, the Energy Technology Support Unit (ETSU).[2] The Oil Shock led to a great deal of economic turbulence, with job stagnation and inflation occurring at the same time ("stagflation"), a new challenge for policy makers.

Writing in 1979, two US scientists noted:

During the years 1975–1978 concern over the increase of CO_2 in the atmosphere expanded from the laboratory into the public policy arena. This was a period during which a profusion of international symposia, technical papers, and public-policy-oriented discussions drew wide attention to the potential dangers of unchecked growth of atmospheric CO_2 and man's alterations of the global carbon cycle.

(Marland & Rotty, 1979)

This should not be overstated – climate change was not a high priority for busy politicians. Nonetheless, under the Labour government of James Callaghan, an interdepartmental committee was created to study the issue and produce a report. It was ready for publication in early 1979 but was not released simply because an election was pending. The Conservative Party, under Margaret Thatcher, won that election. There was discussion within her cabinet about whether the report should even be released (it was equivocal about the likelihood of problems from climate change and silent on carbon capture and storage). In the end, it was released, to no great influence, in February 1980 (Agar, 2015).

Margaret Thatcher was briefed about the issue by her Chief Scientific Advisor John Ashworth. At her first G7 meeting, in Japan, she told a reporter that people should "also be worried about the effect of constantly burning more coal and oil because that can create a band of carbon dioxide round the earth which could itself have very damaging ecological effects" (Nulman, 2016).

3.2.3 Policy stream (specific interest in CCS)

There were four events of note during this period. The first was the earliest example of pushing carbon dioxide underground in order to retrieve oil. While there had been some small-scale pilot carbon dioxide injection projects in the 1960s (Melzer, 2020), the first large-scale use of the technique began on 26 January 1972, when carbon dioxide was injected into an oil field in order to flush out more oil. This initiative was by the oil company Chevron, in Scurry County Texas (Kane, 1979). Ma et al. (2022) note that more than 175 million tonnes of natural CO_2 were injected during the period 1972–2009.

Second, in 1977, a paper titled "On geoengineering and the CO_2 problem" was published (Marchetti, 1977). Marchetti, an Italian physicist, had been involved in discussions, hosted by the International Institute for Advanced Studies Analysis (IASSA), about what responses to the issue of carbon dioxide build-up might be made.[3] Marchetti suggested that it could be captured and stored in either the deep oceans or aquifers.

Third, scientists concerned about climate change began to investigate environmental control technologies (see, e.g., Albanese & Steinberg, 1979) and hold international workshops about what technologies might be available (see, e.g., the International Workshop on the Energy/Climate Interactions, March 3–7, 1980, Munster, Germany). Hoffert et al. (1979) had already discussed some of the implications of Marchetti's proposal.

Fourth, despite the lack of political awareness, British figures in responsible positions began to investigate what might be done about climate change. The Central Electricity Generating Board (CEGB) was concerned enough about the issue to appoint a full-time research officer in 1979

> to keep a watching brief on the scientific aspects of fossil fuel carbon dioxide emissions. Dr Peter S. Liss of the University of East Anglia was engaged as a part-time consultant in the Generation Studies Branch to assess the implications of the greenhouse effect from an industrial perspective.
>
> (Sheail, 1991: 269–270)

The CEGB at this stage did not even mention the possibility of carbon capture and storage, perhaps unsurprisingly, given how speculative the idea was, compared to other more obvious actions, such as nuclear, the expansion of renewables and afforestation.

This detailed account of the first period has hopefully demonstrated that awareness of carbon dioxide build-up – and "technological fixes" for it – has been present for more decades than most people realise.

The streams have not really come together, and indeed, it would be another eight years before a policy window opened for climate change. It is this period which is briefly discussed in the following section.

3.3   Period 2: 1980 to 1989

This section covers the emergence of the climate issue from scientific concern through to the policymaker awareness/acknowledgement stage. Interestingly, through this period, CCS received little research attention. Suggested reasons for this will be discussed in the final paragraphs.

When this period began, the atmospheric concentration of carbon dioxide was 339 ppm, with humans emitting an estimated 19.62 billion metric tonnes. When it ended the concentration was 353 ppm, with emissions at 22.41 billion metric tonnes; the taps were opening, and the bathtub's water level was increasing.

3.3.1   *Problem stream*

From 1980 to 1988, the issue of climate change would appear occasionally in newspapers and on television screens. For example, in December 1981, a documentary, *Warming Warning* – quite probably the first ever entirely devoted to the issue of carbon dioxide build up and its consequences – was broadcast in the UK (Carbon Brief, 2017). Quality newspapers such as *The Guardian* and *The Times* covered the issue. The latter paper ran opinion pieces by the UK diplomat Crispin Tickell, who had been alerted to the issue and had written a book on the subject in the late 1970s. In 1984, in response to a new report by the RCEP on the subject of atmospheric pollution, *The Times* ran a largely supportive editorial, signalling that carbon dioxide build-up needed to be considered as a distinct possibility (The Times, 1984).

In 1985, a crucial meeting of scientists took place in the city of Villach, Austria. At that meeting, scientists realised that the other greenhouse gases they had been tracking, besides carbon dioxide, were also, when added together, significant, and that changes could be expected sooner than had been anticipated. The scientists then began to alert sympathetic politicians (Agrawala, 1998).

As mentioned in Chapter 1, the key year was 1988. Thanks to a very hot summer in the United States, during a presidential election year, the problem became an issue, thanks to determined scientists and environmentalists (Revkin, 1988). *The Changing Environment*, a conference held in June 1988 in Toronto, Canada, was pivotal in putting forward the idea (largely rejected at the time by governments) that Western nations should adopt an emissions reduction target.

3.3.2   *Politics stream*

Prime Minister Margaret Thatcher had displayed indifference or hostility to environmental concerns – especially around the question of acid rain. This made her all the more unexpected as a "policy entrepreneur". Her speech to the Royal Society in late September 1988 coupled the streams and opened a policy window. "The Greenhouse Effect" became a hot political issue, with parties and pressure groups trying to raise ambition and push through their own favoured solutions. The outcomes of the policy window will be discussed more fully in the next section.

3.3.3 Policy stream

Carbon Capture and Storage received remarkably little attention during this period. As US-based commentators noted:

> In the US, preliminary studies were conducted at Brookhaven National Laboratory (Albanese and Steinberg, 1980; Steinberg, 1984). However, it was not until almost 1990 that significant research efforts were undertaken in this field.
>
> (Herzog et al., 1997: 8)

This is borne out by the documentary record in the UK. In 1978, the Labour government had created a *Commission on Energy and the Environment* to "advise on the interaction between Energy Policy and the Environment". Its 1981 publication *Coal and Environment*, seemingly influenced by a CEGB submission, argues:

> No practicable method of controlling carbon dioxide emission from combustion is likely to be possible on a very large scale, because the essence of the combustion of fossil fuel is the conversion of carbon compounds into carbon dioxide. Methods of dissolving it in the ocean have been considered but none seems to be very practicable. If it were deemed necessary to combat the rise in atmospheric levels, the most feasible action would entail burning less fossil fuels, increasing afforestation or increasing the use of nuclear or renewable sources of power.
>
> (Commission on Energy and the Environment, 1981: 176)

This was not the end, however. Gibbins and Lucquiaud (2022) point out that UK studies began with "a study on using post-combustion carbon dioxide (CO_2) capture with amines on coal power plants to supply CO_2 for enhanced oil recovery", citing Roberts (1983). By 1988, the Dutch Ministry of the Environment initiated research into the possibilities of CCS. According to a speech given in 1992:

> The first preliminary studies performed in the late 1980's showed that there were no fundamental technical problems. CO_2 can be removed from flue gases from conventional fossil fuel power plants as well as from coal or oil gasification plants.
>
> (Alders, 1992; see also Hudson, 2015)

The Brookhaven work did not go unnoticed in the UK. In March 1989 an Energy Select Committee report was released (Energy Committee, 1989). It noted that a representative of the CEGB (which in April 1988 had increased its research effort on the Greenhouse Effect – see Highfield, 1988) had told committee members that although the technology was still at a "speculative stage", CO_2 could be captured and stored in the ocean or under the sea-bed (Donovan & Radford, 1989; see also Cookson, 1989).

That same month a major European coal association released a report titled *European Coal and the Greenhouse Effect* (CEPCEO, 1989), which stated that while "limited research has demonstrated the feasibility of removal systems . . . economic and scale up issues have not been adequately researched", and called for further research (funded by taxpayers) to "ascertain its feasibility as a possible remedial strategy". Similarly, presenting at an IEA/OECD Expert Seminar on Energy Technologies for Reducing Emissions of Greenhouse Gases in Paris from 12–14 April 1989, Blok et al. (1989) concluded that *"methods like energy conservation to reduce the carbon dioxide emission and wood planting to compensate for carbon dioxide are cheaper to implement than carbon dioxide recovery"*.

April also saw a presentation to a special day-long Cabinet Meeting organised by Margaret Thatcher on the question of the greenhouse effect in April 1989. At it, a representative of ETSU, the unit set up in the aftermath of the Oil Shock, included CCS as a possible option, albeit with heavy caveats. As number 5 of 8 options (alongside reforestation, energy from waste, nuclear power and fuel substitution), the author described CCS in the following terms:

> 5. The removal of CO_2 from power station flues, so that it is not emitted into the atmosphere. This is thought to be technically feasible and the CO_2 could be pumped down oil wells for enhanced oil recovery. But, it would probably double the cost of electricity production, and it is not yet proven, so I have assumed that there will be no more than perhaps one commercial demonstration by 2020, contributing 2% of our target.
>
> (Currie, 1989: 5)

In this first exuberant period of concern about climate change, many technological fixes were proposed, including a personal favourite (in terms of its outlandishness), the idea of mirrors in space which would reduce the amount of sunlight hitting the earth's surface (Anon, 1989).

3.3.4   Why did the dog not bark?

A good history will point not just to things that *did* happen but to things that did not happen at times when they might have been expected to. This is captured well in the classic passage from the Sherlock Holmes story *The*

Adventure of the Silver Blaze (Doyle, 1894) that captures this, where Holmes speaks of "the curious incident of the dog in the night-time". When the hapless policeman Lestrade replies that "The dog did nothing in the night-time", Holmes replies "That was the curious incident".

A reader in the third decade of the 21st century, with climate change seemingly omnipresent in the news headlines, might wonder why CCS was not "on the agenda" much earlier. There are both general and specific possible answers. Generally, it is important to remember that the confidence of the small group of scientists studying carbon dioxide build-up that there was trouble ahead, and sooner rather than later, was not shared by politicians and industry. Many still thought of it as – in the words of Fairley in 1969 – "science fiction". Perhaps carbon emissions would plateau, or some as yet unstudied natural processes would reduce the impact of any build-up. After all, in 1981 the *Commission on Energy and the Environment*, had claimed that "the nature of the problem is such that they are unlikely to produce definite conclusions for many years. At this stage, we consider that it would be premature to do more than note the potential importance of the issue and that research is being carried out to clarify the problems" (op cit).

Crucially, if there might not be a problem, why spend time, money and energy thinking about enormously expensive and impractical-seeming technological change for what might be a non-problem.

More specifically, there were two factors in the UK in the 1980s that may have further reduced policymaker interest. First, the publicly owned gas infrastructure had been sold off against a general backdrop of privatisations of public assets. Given that CCS would require enormous investment in new infrastructure, at a time when public ownership of even the existing infrastructure was being abolished, it will not have been salient. Second, in the mid-1980s the Thatcher government had defeated the coal miners in a prolonged battle over the future of British coal-mining. In that context, it is easy to understand why so little attention was given to CCS. It was, after all, a technology which, if it succeeded, might prolong the life of coal-mining.

3.4   Period 3: 1990–1997

In this period, concern about climate change and the investigation of a particular new potential technofix (CCS) began to come together, thanks to a small number of organisations determined to investigate what might be possible, including both the British state and coal industry actors.

When this period began, the atmospheric concentration of carbon dioxide was 354.2 ppm, with an estimated annual amount of carbon dioxide being emitted by humans at 22.76 billion tonnes. At the end of the period, the concentration was 363.5 ppm, with emissions at 24.3 billion tonnes.

3.4.1   Problem stream

The Intergovernmental Panel on Climate Change (IPCC) was formed in late 1988. In an impressively quick time it produced its first Assessment Report, delivered as three Working Group reports and then a synthesis report by mid-1990.

The science of climate change was contested by groups like the George Marshall Institute and the Global Climate Coalition (see Oreskes & Conway, 2010; Gelbspan, 1998, 2004). Their publications were given significant (and often uncritical) coverage in the business press and "right-wing" media outlets. However, there was enough agreement that the "enhanced Greenhouse Effect" was real, and with significant proportions of the broader public alarmed and urging action, the negotiations towards a global treaty began (see Paterson, 1996; Legget, 2001, for eyewitness accounts).

Coal was "in the frame", and coal industry figures pondered the future (see Harrison, 1990). In April of 1991, the World Coal Institute held a conference on the question of "Coal in the Environment" in London (Rubin, 1991).

The IPCC produced a second, supplemental, report for the negotiations in 1992 and a second Assessment Report in 1995. This report said human activities, namely the emissions of fossil fuels and deforestation, were having a "discernible impact". It – and its authors – was fiercely attacked by the Global Climate Coalition and other groups, but the science was robust.

3.4.2   Politics stream

Internationally, the main event in this period was the negotiations towards a global climate treaty. The United States, under the Presidency of George H. W. Bush first tried to say that further research was required and that it was premature to begin negotiations towards a treaty. When that position became untenable, it then stated that targets and timetables for reductions of emissions by wealthy countries could not be part of the treaty text. The proponents of targets and timetables, fearing that the United States would not attend the Earth Summit in Rio 1992 at which the treaty was to be opened for signatures, withdrew the proposal. Alongside resistance from the United States and various oil-producing states, there was also significant resistance to a strong treaty from oil and gas companies, automakers and other groups that calculated they would lose out if emissions targets were set (Leggett, 2001).

The treaty – the UNFCCC – had as its objective the following.

> The ultimate objective of this Convention and any related legal instruments that the Conference of the Parties may adopt is to achieve, in accordance with the relevant provisions of the Convention, *stabilization of greenhouse gas concentrations in the atmosphere at a level that would prevent dangerous anthropogenic interference with the climate system.* Such a level should be achieved within a time frame sufficient to allow

ecosystems to adapt naturally to climate change, to ensure that food production is not threatened and to enable economic development to proceed in a sustainable manner (emphasis added).

(UNFCCC, 1992)

It should be clear to readers in the third decade of the 21st century that this objective has not been met. Whatever CCS achieves, in terms of reducing the acceleration of atmospheric concentrations of carbon dioxide (the water in the bathtub), there is already water splashing over the sides.

The question of emissions cuts for rich countries was not in the Treaty, but that did not mean it was no longer an issue. The main outcome of COP1, held in March–April 1995, was the "Berlin Mandate", which called for developed countries to come to the third COP with agreements for emissions reductions. That meeting was held in Kyoto, Japan, and gave rise to the Kyoto Protocol.

Although the European Community (now the European Union (EU)) was seen to have played a relatively constructive role in the negotiations, pushing for increased ambition, it failed – thanks to concerted opposition from business lobbies – to institute a carbon tax. One European nation that did create such a tax though was Norway, with direct consequences for CCS, since it was the impetus for the state-owned oil company Statoil (since renamed Equinor) to begin storing carbon dioxide under the North Sea in 1996.

Nationally, the UK's climate ambitions were "in the middle of the pack". The Thatcher government strenuously rejected the idea of steep emission cuts but instead committed to stabilising emissions at 1990 levels by the year 2000. This, in the context of ongoing use of coal for electricity (albeit with more gas entering the energy system during the so-called "Dash for Gas"), meant that – alongside proposed energy efficiency measures – more would need to be done. This helped create interest in CCS.

3.4.3 Policy stream

The interest in CCS for helping prolong coal's longevity came, predictably enough, from the British Coal and the Department of Energy (the former was a rebranding of the National Coal Board, responsible for the mining of coal in the UK). An initiative by these two met with success. After lobbying from them, the International Energy Agency's Ministerial Communiqué in 1991 included agreement on research into the "*integration of energy and environmental goals, particularly in such areas as renewables, nuclear power systems, innovative conservation technologies, CO$_2$ capture and utilization, and fossil utilization*" (emphasis added) (Scott, 1993: 58). Park (1991) undertook "A preliminary estimate of the capacity of UK oil and gas fields for the storage of carbon dioxide" for British Coal.

However, both the CEGB and British Coal were being prepared for privatisation, and as many warned at the time, this was to have an extremely deleterious impact on the amount of research conducted into "clean coal", including carbon sequestration (Tieman, 1991; Nuttall & Hawkes, 1992; Harrison & Williams, 1993).

The year 1991 saw the formation of an International Energy Agency Greenhouse Gas Programme (IEAGHG) (Jack et al., 1992). In the same year, British Coal also created an in-house programme at its long-standing Coal Research Establishment (CRE) to support the IEAGHG work (Bower et al., 1992: 224; Webster et al., 1993). This work centred on "investigating options for removing carbon dioxide from power stations in case expensive fallback options become necessary" (Bower et al., 1992). Early conclusions were that "C02 control could be retrofitted to existing pulverised coal plants, but gasification based systems are more promising" (op. cit). In an early example of how CCS and hydrogen hopes are entwined, it suggested that such actions "would permit a clean hydrogen fuel to be fired in a gas turbine and involve removal of 90% of the C02" (Bower et al., 1992: 222).

The IEAGHG was officially launched on 20 November 1991, with the involvement of a number of countries, including Australia, Denmark, the Netherlands, the United States and the United Kingdom (Jack et al., 1992: 813). Its mission statement will seem familiar, given that variations on it have been repeated ever since.

- Evaluate on a full fuel cycle basis the technologies used for the abatement, control, use and disposal of CO_2 and other greenhouse gases derived from fossil fuel use. The technical feasibility, performance, environmental benefits and impacts of the technologies will also be studied.
- Estimate the impact the implementation of the technologies would have on the economy and energy markets.
- Disseminate the results of the programme activities to the participants and prepare R&D proposals for the favoured technical options.
- Sponsor or conduct collaborative R&D projects in the fields of CO_2 removal, utilisation, transport and disposal.

As these discussions were taking place, individual oil companies such as Imperial Oil, a Canadian subsidiary of Exxon, were investigating underground disposal of CO_2. An April 1991 document, according to Climate Files (n.d.), found that "*such an undertaking does not* "*appear to be economic*" *and would* "*achieve a relatively minor impact in reducing CO$_2$ emissions*", even with significant collaboration among stakeholders*".

In March 1992 the first gathering of the international research community investigating CO_2 control technologies, the First International Conference on Carbon Dioxide Removal, took place in the Netherlands, becoming a biannual event (Blok et al., 1992; Herzog et al., 1997: 8).

Meanwhile, the IEAGHG, and that of the other British Coal initiatives, spawned a significant amount of research into various aspects of CCS. On storage, there was early work by the British Geological Survey (Holliday et al., 1991; Holloway et al., 1996). There was optimism about the

[u]se of disused oil and gas fields to store CO: is attractive, not only because of the conceptual tidiness of using the space left by removal of hydrocarbons, but also because the knowledge gained about such fields from their previous use (e.g. detailed geology, existence of a geological seal) gives confidence in their use for storing CO_2.

(Freund & Ormerod, 1997: S200)

They noted that *"significant legal and jurisdictional issues may also have to be faced over use of the deep ocean for storage of CO_2"*. Flagging to advocates of ocean disposal that there might be trouble ahead, they noted that *"legislators concerned with use of the oceans for waste disposal (e.g. 1972 London Agreement) are responding to international pressure to protect the deep oceans and are progressively tightening the legislation"* (op cit. S202)

However, there was also realism about the very significant costs that would be involved, especially with capture: *"Analysis shows that the cost of CO_2 capture dominates these added figures unless CO, transport distances are very large"* (Riemer & Ormerod, 1995).

The second half of this period also saw an uptick in studies and public-facing explanations (Herzog & Drake, 1996; Herzog et al., 1997; Herzog et al., 2000).

The most noteworthy event during this period was not the work of academics[4] but rather an event in the real world which is still pointed to by CCS advocates as proof of concept. In September 1996, the Norwegian state oil company Statoil began to inject CO_2 into the Sleipner West field under the North Sea (Furre et al., 2017). As a Massachusetts Institute of Technology website states, with disarming candour, "The Sleipner CO_2 gas processing and capture unit was built in order to evade the 1991 Norwegian CO_2 tax. Sleipner obtains CO_2 credit for the injected CO_2 and does not pay the tax" (Carbon Sequestration Initiative, n.d).

3.5　Period 4: 1998–2002

This period shows the streams filling, and increasing political and economic interest in CCS, as part of a more general effort to export environmental control technology to developing countries. One of the astonishing events of the 1990s had been the endless economic growth of China – fuelled largely by coal. The UK government, and UK industry, saw opportunities ahead.

When this period began, the atmospheric concentration of carbon dioxide was 366 ppm, with an estimated annual amount of carbon dioxide being emitted by humans at 24.21 billion tonnes. At the end of the period the concentration was 373 ppm, with emissions at 26.28 tonnes.

3.5.1   Problem stream

Three particular events stand out during this time. First, in August, the RCEP, created by Harold Wilson in late 1969, released its 22nd report, titled *Energy – the Changing Climate*. It called for the UK government to raise its ambition on emissions reductions, calling for a target of a 60% reduction by 2050 (this was considered radical for its time, long before "net zero" was a household phrase). This report, from an impeccably small-c conservative organisation, made the job of those pushing for increased action easier. Second, shortly after, Western Europe, and especially the UK, was hit by severe flooding, which campaigners said was related to climate change. The third event was the release of the IPCC's Third Assessment Report (IPCC, 2001).

3.5.2   Politics stream

Internationally, this period saw the agreement of the Kyoto Protocol, with developed nations committing to (relatively small) emissions cuts over the coming decade. Although there was doubt that the United States would ever ratify it (and indeed, President George W. Bush withdrew the United States from the negotiations in March 2001), it nonetheless raised interest in not only carbon trading, and so-called "Joint Implementation" (paying other nations to make emissions reductions because that would be cheaper and more convenient than doing it at home), but also in technologies that might help reduce carbon dioxide levels. For example, as late as 1997 the International Petroleum Industry Environmental Conservation Association (IPIECA) released a book about climate change with no mention of CCS (Flannery & Clarke, 1997).[5] After that date, CCS was no longer absent from such volumes.

Nationally, the UK government tried to introduce a carbon pricing mechanism, but the proposal was watered down, thanks to fierce and successful business lobbying. Meanwhile, UK emissions were no longer falling as quickly as they had (the previous reductions had been down to a steady decline in coal in the UK energy mix and offshoring of some heavy industry). Without further action, even existing targets, let alone what the RCEP were advocating, would soon be out of reach.

3.5.3   Policy stream

CCS was gaining attention, but there was still a great deal of scepticism. This is nicely captured in the following quote by a US-based engineer, with an unconscious echo of the Fairley reference to science fiction in 1969.

> The idea of capturing and storing billions of tons of CO_2 for long spans of time is still regarded by many as pure fantasy and science fiction. When I started working on C02 sequestration about 10 years ago, many regarded this concept as the craziest idea they ever heard.
>
> (Bergman, 1999)

Science fiction or not, the concept started to gain traction, thanks in part to the efforts of the "C02 Capture Project", a group of major oil companies – including BP, Chevron and Petrobas, Eni, Shell and Statoil, with government funding from the United States, the EU and Norway (Wright et al., 2004).

Academics were active within the newly created Tyndall Centre (their work will be discussed more in Chapter 4). Groups such as ETSU continued to work with the Department of Trade and Industry (DTI) and sections of the IEA to produce pamphlets aimed at raising awareness of the technology with relevant industry and political stakeholders (ETSU, 1999), still arguing that ocean disposal might be both legally and technically possible.

Advocates of the technology were making use of arguments from adjacent policies. The Blair government, first elected in May 1997, had produced a Competitiveness White Paper with the predictable title *Our Competitive Future: Building the Knowledge-Driven Economy.* (HMG, 1997). The authors of an Energy Paper 67 on Cleaner Coal Technologies argued that

British business must compete by exploiting capabilities which competitors cannot easily match or imitate- knowledge, skills and creativity. . . . In the power sector, the major components of conventional power plant are increasingly being supplied by the home market in developing countries such as China and India rather than by overseas companies. Such markets will only remain open to UK industry if it can continue to supply more advanced components which offer improvements in efficiency, environmental performance and innovation etc.

(DTI, 1999)[6]

Continuing this, DTI commissioned a report on the question of transfer of cleaner coal technologies to China (Watson et al., 2000).

In 2001, the businessman Jeff Chapman, who would later be the first head of the CCSA, established an industry interest group with support from the UK Trade and Investment (a state body that supports British businesses to find export opportunities). This became the foundation of the CCSA, discussed in the next chapter (Powerbase, nd).

In August 2001, the Department of Trade and Industry launched a consultation paper – *Review of the Case for Government Support for Commercial Scale Cleaner Coal Demonstration Plant* (DTI, 2001), with CCS as one of the technologies referenced.

In February 2002, the Performance and Innovation Unit, set up by Prime Minister Tony Blair, released a report (PIU, 2002) that further drove CCS up the agenda (Jordan, 2002; MacKerron, 2009).

Finally, the threads came together. On 17 September 2002, the Labour Minister for Energy, Brian Wilson, announced a study with the following objectives:

• Establish the technical feasibility of CO_2 capture and storage as a low-carbon option.

- Define the potential technical, market, economic, public acceptability and legal barriers, and consider options for their solution.
- Establish the circumstances that could make the option competitive with other abatement measures.
- Consider the size of the potential contribution to UK abatement targets.
- Assess export opportunities for the technology.
- Define the role of the government in taking forward CO_2 capture and storage.

The report would be released the following year, as the *Review of the Feasibility of Carbon Dioxide Capture and Storage in the UK* (DTI, 2003). CCS had officially arrived.

3.6   Conclusion

This chapter has covered 30 years of the gradual rise of awareness of the threat of climate change. It moved from "science fiction" to accepted fact. Meanwhile, a speculative idea around capture and ocean storage was – especially from the early 1990s – taken more and more seriously. During this period, the costs of CCS were acknowledged – see, for example, Topper et al. (1992) – but were not considered a hindrance to further research and advocacy. That would change in the next period and has remained so ever since.

Notes

1   This bipartisanship – which can be overstated – has been an interesting feature of the British scene for the last 50 years. Whether actions by the Conservative government in July–September 2023 represent a fundamental that remains to be seen – see Hudson, 2023, for further discussion.

2   "The Energy Technology Support Unit was brought formally into existence in mid-April [1974] following an exchange of letters between the Department of Energy and the Atomic Energy Authority, as a result of negotiations which took place over some months previously. The stimulus was the crisis precipitated by the curtailment of oil production and the rapid escalation in fuel prices during the second half of 1973" (Dawson, 1974).

3   The International Institute for Applied Systems Analysis (IIASA) was launched in the early 1970s as a bridge-building body for scientists in the Soviet Union and the US/Europe. IIASA was also a key location for early conversations about Bio-Energy Carbon Capture and Storage discussed in Chapter 5.

4   Is it ever?

5   The book does however mention Miller Field, which will be discussed in Chapter 4.

6   This is telling – even when "industrial policy" is not "allowed" to be spoken of in polite company, for fear of attracting criticism for "mercantilism" and "picking winners", it nonetheless continues to happen.

References

Agar, J. 2015. 'Future forecast – Changeable and probably getting worse': The UK Government's early response to anthropogenic climate change. *Twentieth Century British History*, 26(4), 602–628.

Agrawala, S. 1998. Context and early origins of the intergovernmental panel on climate change. *Climatic Change*, 39(4), 605–620.

Albanese, A. S. and Steinberg, M. 1979. *Environmental Control Technology for Atmospheric Carbon Dioxide*. New York: Brookhaven National Laboratory, US Department of Energy.

Albanese, A. S. and Steinberg, M. 1980. Environmental control technology for atmospheric carbon dioxide. *Energy*, 5(7), 641–664. https://doi.org/10.1016/0360-5442(80)90044-4

Alders, J. G. M. 1992. Opening speech on the occasion of the first international conference on carbon dioxide removal. *Energy Conversation and Management*, 33(5–8), 283–286.

Anon. 1989. Space shield plan to cut sunlight. *New Zealand Herald*, August 17. https://allouryesterdays.info/2023/08/16/august-17-1989-space-shields-to-save-the-earth/

Bergman, P. 1999. Geological sequestration of C02: A status report. In *Energy Conversation. Management*, edited by Eliasson, B., Riemer, P. and Wokaun, A. X. Amsterdam: Elsevier, pp. 169–173.

Blok, K., Hendriks, C. and Turkenburg, W. 1989. *The Role of Carbon Dioxide Removal in the Reduction of the Greenhouse Effect*. Utrecht: The Department of Science & Technology, University of Utrecht.

Blok, K., Turkenburg, W. C., Hendriks, C. A. and Steinberg, M. (eds). 1992. *Proceedings of the First International Conference on Carbon Dioxide Removal*. Oxford: Pergamon Press, p. 544.

Bower, C., Steve, G., Iain, S. and Geoff, F. 1992. C02 removal as a fall back option for power generation? *Energy & Environment*, 3(3), 222–237.

Broecker, W. 1975. Climatic change: Are we on the brink of a pronounced global warming? *Science*, 189(4201), 460–463. https://doi.org/10.1126/science.189.4201.460

Carbon Brief. 2017. *The 1981 TV Documentary that Warned about Global Warming*. www.carbonbrief.org/warming-warning-1981-tv-documentary-warned-climate-change/

Carbon Sequestration Initiative. n.d. *Sleipner Fact Sheet*. https://sequestration.mit.edu/tools/projects/sleipner.html

CEPCEO. 1989. *European Coal and The Greenhouse Effect*. Comité d'Étude des Producteurs de Charbon d'Europe Occidentale CEPCEO.

Climate Files. n.d. *1991 Imperial Oil Discussion Paper on Underground Disposal of Carbon Dioxide*. www.climatefiles.com/exxonmobil/1991-imperial-oil-discussion-paper-on-underground-disposal-of-carbon-dioxide/

Commission on Energy and the Environment. 1981. *Coal and Environment*. London: HMSO.

Cookson, C. 1989. The burning questions about fossil fuels. *Financial Times*, March 9, p. 24.

Currie, K. 1989. *Options for Mitigating the Greenhouse Effect*. Energy Technology Support Unit (ETSU-R-54).

Dawson, J. K. 1974. The energy technology support unit at harwell. *Physics Bulletin*, 25, p. 381

Donovan, P. and Radford, T. 1989. Sea dumps 'could cut electricity pollution'. *The Guardian*, March 7, p. 4.

Doyle, A. C. 1894. *The Memoirs of Sherlock Holmes*. London: G. Newnes Ltd.

DTI. 1999. *Energy Paper 67*. London: HMSO.

DTI. 2001. *Review of the Case for Government Support for Commercial Scale Cleaner Coal Demonstration Plant*. London: HMSO.

DTI. 2003. *Review of the Feasibility of Carbon Dioxide Capture and Storage in the UK*. London: HMSO.

Energy Committee. 1989. *Energy Policy Implications of the Greenhouse Effect*. London: HMSO.

ETSU. 1999. *Cleaner Coal Technologies: Options*. London: ETSU.

Fairley, A. 1969. Information: London's drowning. *The Listener*, April 30, p. 475.

Flannery, B. P. and Clarke, R. (eds). 1997. *Global Climate Change: A Petroleum Industry Perspective*. London: IPIECA.

Freund, P. and Ormerod, W. G. 1997. Progress toward storage of Carbon Dioxide. *Energy Conversion Management*, 38(Suppl.), S199–S204.

Furre, A. K., Eiken, O., Alnes, H. M., Vevatne, J., Nesland, K. and Anders, F. 2017. 20 years of monitoring CO_2-injection at Sleipner. *Energy Procedia*, 114, 3916–3926. https://doi.org/10.1016/j.egypro.2017.03.1523

Gelbspan, R. 1998. *The Heat Is on. The Climate Crisis, the Cover-Up, the Prescription*. New York: Perseus Books Group.

Gelbspan, R. 2004. *Boiling point: How Politicians, Big Oil and Coal, Journalists and Activists Are Fueling the Climate Crisis – and What We Can Do to Avert Disaster*. New York: Basic Books.

Gibbins, J. and Lucquiaud, M. 2022. The development of UK CCUS strategy and current plans for large-scale deployment of this technology. *Annales des Mines – Responsabilité et Environnement*, 105(1), 26–30.

Hansard. 1971. *North Atlantic Dumping of Chemical Waste*, July 21. https://hansard.parliament.uk/commons/1971-07-21/debates/59b70e58-e195-4b05-827e-db0a35795e34/NorthAtlantic(DumpingOfChemicalWaste)

Harrison, J. 1990. Global warming: the coal industry view. *New Scientist*, 127(1732).

Harrison, J. and Williams, A. 1993. Coal privatisation threat to research. *The Times*, March 3, p. 15.

Herzog, H. and Drake, E. M. 1996. Carbon dioxide recovery and disposal from large energy systems. *Annual Review of Energy and Environment*, 21, 145–166.

Herzog, H., Eliasson, B. and Kaarstad, O. 2000. Capturing greenhouse gases. *Scientific American*, 282(2), 72–79.

Herzog, H., Elisabeth Drake, E. and Adams, E. 1997. *CO Capture, Reuse, and Storage Technologies 2 for Mitigating Global Climate Change A White Paper Final Report DOE Order No. DE-AF22–96PC01257*. US Department of Energy.

Highfield, R. 1988. £1.25m boost for 'greenhouse effect' research. *Daily Telegraph*, April 22, p. 8.

HMG. 1997. *Our Competitive Future: Building the Knowledge-Driven Economy*. London: HMSO

Hoffert, M. I., Wey, Y. C., Callegari, A. J. and Broecker, W. C. 1979. Atmospheric response to deep-sea injections of fossil-fuel carbon dioxide. *Climatic Change*, 2(1), 53–68.

Holliday, D. W., Williams, G. M., Holloway, S., Savage, D. and Bannon, M. P. 1991. *A Preliminary Feasibility Study for the Underground Disposal of Carbon Dioxide in UK*. Nottingham: British Geological Survey, p. 32. (WE/91/020) (Unpublished)

Holloway, S., Heederick, J. P., van der Mcer, L. G. H., Czernichowski-Lauriol, I., Harrison, R., Lindeberg, E., Summerfield, I. 1L Rochelle, C., Schwarzkopf, T., Kaarstad, O. and Berger, B. 1996. *The Underground Disposal of Carbon Dioxide – Summary Report* (European Communities Joule 11 Programme). Keyworth, Nottingham: British Geological Survey.

Hudson, M. 2015. "Dutch citizens sue over climate change". A little wade down memory canal. #climatehistory. *Marchudson.net*. https://marchudson.net/2015/04/15/dutch-citizens-sue-over-climate-change-a-little-wade-down-memory-canal-climatehistory/#more-376

Hudson, M. 2023. Rishi Sunak's green backtracking contrasts strongly with previous prime ministers' efforts. *The Conversation*, August 3. https://theconversation.com/rishi-sunaks-green-backtracking-contrasts-strongly-with-previous-prime-ministers-efforts-210917

IPCC. 2001. *Third Assessment Report*. https://www.ipcc.ch/assessment-report/ar3/

Jack, A. R., Audus, H. and Riemer, P. W. F. 1992. The IEA greenhouse gas R&D programme. *Energy Conversion and Management*, 33(5–8), 813–818. https://doi.org/10.1016/0196-8904(92)90088-e

Jordan, A. 2002. Decarbonising the UK: A 'Radical Agenda' from the UK Cabinet Office. *The Political Quarterly*, 73(3), 344–352.

Kane, A. V. 1979. Performance review of a large-scale CO_2-WAG enhanced recovery project, SACROC unit kelly-snyder field. *Journal of Petroleum Technology*, 31(02), 217–231. https://doi.org/10.2118/7091-pa

Kennet, L. 1968. *Pollution in the Human Environment: Proposals to Set up a Committee or Other Body to Undertake a Study (1968–69)*, Minute from Lord Kennet to Minister of Housing and Local Government, 15 July 1968 TNA: HLG 127/1193

Leggett, J. K. 2001. *The Carbon War: Global Warming and the End of the Oil Era*. London: Psychology Press.

Ma, J., Li, L., Wang, H., Du, Y., Ma, J., Zhang, X. and Wang, Z. 2022. Carbon capture and storage: History and the road ahead. *Engineering*, 14, 33–43.

MacKerron, G. 2009. Lessons from the UK on urgency and legitimacy in energy policymaking. In *Energy for the Future. Energy, Climate and the Environment Series*, edited by Scrase, I. and MacKerron, G. London: Palgrave Macmillan. https://doi.org/10.1057/9780230235441_5

Marchetti, C. 1977. On geoengineering and the CO_2 problem. *Climate Change*, 1, 59–68.

Marland, G. and Rotty, R. M. 1979. Carbon dioxide and climate. *Reviews of Geophysics*, 17(7), 1813. https://doi.org/10.1029/rg017i007p01813

Martin-Nielsen, J. 2018. Computing the climate: When models became political. *Historical Studies in the Natural Sciences*, 48(2), 223–245.
Melzer, S. 2020. A Brief History of CO_2 EOR, New developments and reservoir technologies for CO_2 EOR in conjunction with Carbon Capture, Utilization and Storage (CCUS). In *Presented at the 26th Annual CO_2 Conference Midland*. Midland, TX: CCUS.
National Research Council. 1977. *Energy and Climate: Studies in Geophysics*. Washington DC: National Academies Press.
Nulman, E. 2016. *Climate Change and Social Movements: Civil Society and the Development of National Climate Change Policy*. New York: Springer.
Nuttall, N. and Hawkes, N. 1992. Clean coal 'blocked' by lack of funds. *The Times*, October 16, p. 2.
Oreskes, N. and Conway, E. 2010. *The Merchants of Doubt*. London: Bloomsbury.
Park, R. S. 1991. *A Preliminary Estimate of the Capacity of UK Oil and Gas Fields for the Storage of Carbon Dioxide*. Unpublished report for British Coal.
Paterson, M. 1996. *Global Warming and Global Politics*. London: Routledge.
PIU. 2002. *The Energy Review*. London: Cabinet Office.
Powerbase. n.d. Jeff Chapman. https://powerbase.info/index.php/Jeff_Chapman
RCEP. 1971. *First Report of the Royal Commission on Environmental Pollution*. London: HMSO
RCEP. 2000. *Energy – The Changing Climate. 22nd Report*. London: JHMSO.
Revkin, A. C. 1988. The endless summer: Living with the greenhouse effect. *Discover Magazine*. www.discovermagazine.com/environment/special-report-endless-summerliving-with-the-greenhouse-effect
Riemer, P. W. F. and Ormerod, W. G. 1995. International Perspectives and the results of carbon dioxide capture disposal and utilisation studies. *Energy Conversion Management*, 36(6–9), 813–818,
Roberts, H. 1983. The logistics and economics of a CO_2 – flood. *Presented at the International Energy Agency Workshop*, 25 August–27 July. Vienna: Oil Recovery Projects Division Report, AEE Winfrith.
Rubin, E. 1991. *Environmental constraints: Threat to Coal's Future? Keynote Session Presentation to the World Coal Institute Conference on Coal In the Environment*. London: World Coal Institute.
Scott, R. 1993. *IEA the First 20 Years. Vol 2 Major Policies and Actions*. Paris: IEA, pp. 57–58.
Sheail, J. 1991. *Power in Trust: The Environmental History of the Central Electricity Generating Board*. Oxford: Clarendon Press.
Sims, P. D. 2016. *The Development of Environmental Politics in Inter-War and Post-War Britain*. PhD Thesis, Queen Mary, University of London. https://qmro.qmul.ac.uk/xmlui/bitstream/handle/123456789/23653/Sims_P_PhD_final_150816.pdf
Steinberg, M. 1984. *An Analysis of Concepts for Controlling Atmospheric Carbon Dioxide, DOE/CH/00016–1*. Brookhaven: Brookhaven National Laboratory.
The Times. 1984. The greenhouse effect. *The Times*, February 23, p. 13.

Tieman, R. 1991. Strategy needed to keep home fires burning on coal. *The Times*, July 18, p. 27.

Topper, J. M., Bower, C. J., Summerfield, I. R. and Hughes, I. S. C. 1992. The British coal global warming R&D programme. *Energy Conversion and Management*, 33(5–8), 803–811. https://doi.org/10.1016/0196-8904(92)90087-d

UNFCCC, 1992. *United Nations Framework Convention on Climate Change*. New York. https://unfccc.int/files/essential_background/background_publications_htmlpdf/application/pdf/conveng.pdf

Watson, J., Oldham, G., Mackerron, G., Thomas, S. and Xue, L. 2000. *The Transfer of Cleaner Coal Technologies to China: A UK Perspective*. London: DTI/SPRU.

Weart, S. R. 2008. *The Discovery of Global Warming: Revised and Expanded Edition*. Harvard: Harvard University Press.

Webster, I. C., Audus, H., Ormerod, W. G. and Riemer, P. W. F. 1993. *International Cooperation in Greenhouse Gas Mitigation in the Second World Coal Institute Conference*. London: World Coal Institute.

Winsbury, R. 1969. Pollution – the cost to industry, the risk to life. *Financial Times*, October 22, p. 15.

Wright, I. W., Lee, A., Middleton, P., Lowe, C., Imbus, S. W. and Miracca, I. 2004. CO_2 Capture project: Initial results. *SPE International Conference on Health, Safety, and Environment*. https://onepetro.org/SPEHSE/proceedings-abstract/04HSE/All-04HSE/SPE-86602-MS/71826

Wright, P. 1970. Pollution catastrophe denied. *The Times*, August 26, p. 1.

4 From 2003 to 2015: "high hopes, repeatedly dashed"

Contents

4.1 Introduction

Twenty-five years after the first idea, CCS had finally arrived at a place of prominence, if not acceptance. This was primarily because the climate issue – despite what persistent denialists were saying – was not a hoax. It was not going to go away. Fossil-fuel energy companies knew that sooner or later they would be under the spotlight and would need plausible answers. Meanwhile, governments that had made bold promises of action were coming to realise that it was one thing to announce "round-number" targets but quite another to create and implement policy that would deliver. This was especially so in the UK, where gas prices were climbing and so coal use was increasing.

Between 2003 and 2005, there was a flurry of interest and public announcements about CCS. In 2005, BP proposed a CCS project, DF1, at Peterhead in Scotland. Eventually, in 2007, BP withdrew the proposal, blaming the Blair government for delays in decision-making about funding. The Brown government then announced a Commercialisation Competition, which ran from late 2007 until late 2011, when the last eligible project pulled out. Shortly after, a new CCS roadmap and a new competition were announced. Then, on 25 November 2015, it all fell apart.

This chapter will first cover the two-year period and the events unfolding within it, before giving detailed chronological accounts of the BP proposal and the first and second competitions. Within the account, it will flag the

DOI: 10.4324/9781003461067-4

arrival of new bodies, such as the CCSA, the Committee on Climate Change (CCC) and the Department of Energy and Climate Change (DECC). It will also address the significance of the Climate Change Act (CCA) and the protest movement against "unabated coal" – the potential of a new movement, not just anti-coal, but anti-CCS, will be discussed in Chapter 6.

4.2   2002–2005: The CCS window opens, both internationally and nationally

There were both international and national factors propelling climate change – and CCS – further up the policy agenda. Geopolitically, the most significant events in this period were the build-up to the illegal invasion of Iraq in 2003, the invasion itself and the occupation. On climate change, the most important events were the difficulties around the Kyoto Protocol. In March 2001, shortly after taking office, President George Bush announced that he was withdrawing the United States from negotiations. This meant that there was a great deal of uncertainty about what the international rules – and markets – would be. The matter was compounded by Australia's withdrawal the following June (despite having been able to negotiate extremely generous terms at the Kyoto meeting). Australia and the United States were then at the forefront of bilateral and international "spoiler" organisations that focussed far more on technological options for addressing climate change. However, finally, in 2005 Russia ratified the Protocol, and it became international law.

The United States was keen to promote technological approaches, and the EU, as it had with other US preferences, acceded. On 5–6 February 2003, shortly before the attack on Iraq, the United States and EU convened their first bilateral "U.S.-EU Joint Meeting on Climate Change Science and Technology Research" in Washington. The meeting led to an agreement on cooperative research activities in six areas: carbon cycle research; aerosol-climate interactions; feedback, water vapour and thermohaline circulation; integrated observation systems and data; CCS; and hydrogen technology and infrastructure.

On CCS, the United States and EU agreed on four actions.

1   Identify potential areas of collaboration on CCS.
2   Foster collaborative research and development projects.
3   Identify opportunities to discuss the perspectives of governments and other key stakeholders.
4   Discuss planning, including research and development, for large integrated sequestration and energy plant projects.

(US Department of State, 2003)

This should be seen in the context of other multilateral efforts that the United States was making, such as initiating a Carbon Sequestration Leadership Forum.[1]

Meanwhile, the IPCC had agreed to produce one of its periodic Special Reports, on the topic of CCS. This report, released in 2005, led much-sought legitimacy to the nascent technology (IPCC, 2005).

Meanwhile, the UK used its presidency of the EU to push through a CCS agreement at a Summit in September 2005. The Near Zero Emission Coal agreement sought to help demonstrate CCS in China and the EU by 2020. There was a further bilateral initiative between the UK and China (Zhongyang et al., 2009).

Finally, beyond the work of the CO_2 Storage Project mentioned in the last chapter, individual oil companies were starting to make noises about CCS. For example, Shell released a 52-page report, full of glossy photos, called *Meeting the Energy Challenge*. There was a short section on "carbon dioxide" capture, noting that it "could take more than a century" to stabilise carbon dioxide emissions. Shell also announced it had that year set up a CO_2 capture team "*with technical and commercial experts from across Shell. Its goal is to dramatically cut the cost of capturing and reusing CO_2, by 2010*" (Shell, 2002: 30; see also Rowell & Stockman, 2021).

Nationally, the Blair government, then in its second term and with a large majority in the House of Commons, had – as is always the case – a set of thorny issues on its plate. One, which would continue to exert a kind of negative influence, was the war in Iraq. At the same time that huge marches were taking place against British involvement, the 2003 Energy White Paper, *Our energy future – creating a low carbon economy*, was released HMG, 2003). This drew on work by the Energy Review, discussed in the last chapter. It has come to be regarded as pivotal and as the first serious attempt to grapple with the pressing need to reduce emissions from energy (the idea of the need for "net zero" was then still the preserve of "eco-loons").

In Chapter 6, it states:

> Our preference is for a market framework with the right regulatory frame-work. But neither should we allow ourselves to become overly dependent on any one fuel source across the whole economy or in a specific sector, such as electricity generation.

Crucially, for CCS, it stated that "the policies we put forward in this paper will encourage the long-term development of new, more diverse and cleaner energy technologies that will promote both energy reliability and our low-carbon objectives".

Noting that about a third of the UK's power output came from coal, it said that "in a low carbon economy the future for coal must lie in cleaner coal technologies – which can increase the efficiency of coal-fired power stations and thereby reduce the amount of carbon they produce – or *carbon capture and storage*".

In paragraph 6.63, the commitment is made:

> We will therefore set up an urgent detailed implementation plan with the developers, generators and the oil companies to establish what needs to be done to get a demonstration project off the ground. This study will reach conclusions within six months to enable firm decisions to be taken on applications for funding from international sources as soon as possible thereafter.

This was indeed accomplished. The September 2003 *Review of the Feasibility of Carbon Dioxide Capture and Storage in the UK*, quoted at the end of Chapter 3, concluded that fossil fuels would continue to be a major source of energy for both the UK and the rest of the world for decades. CCS might help to make radical cuts in these emissions (DTI, 2003).

Further, the DTI had already commissioned researchers at the Tyndall Centre to investigate the public acceptability of CCS (McLachlan, 2003).

Meanwhile, broadsheet newspapers were beginning to run features on the topic (Adam, 2003; Arjuna, 2004), laying out the numerous challenges the technology faced.

Additionally, academic networks were forming. In a letter to the *Independent* titled "New Weapon Against Global Warming", a group of academics stated:

> Carbon storage has also lacked the well-organised industrial and establishment backing that nuclear power has enjoyed. Fortunately the Energy White Paper recognises the potential for carbon storage and the UK Research Councils, through the newly-established UK Energy Research Centre, will shortly be setting up a wide-reaching stakeholder network. This will help raise the profile of CCS as an element in future energy debates.
>
> (Ali et al., 2004)

The following year a public consultation was run against the backdrop of a report by the research group Cambridge Econometrics, which signalled that the UK was set to miss its 2010 target of a 20% reduction in carbon dioxide emissions by approximately 8% (Anon, 2004, European Daily Electricity Markets).

In the aftermath of the attack on Iraq proving politically difficult – there were, it emerged, no Weapons of Mass Destruction to be found – and with the UK hosting the G7 meeting in 2005, alternative subjects for discussion were valued. Two emerged – one, aid and poverty ("Make Poverty History"), and the other, climate change.

In late 2004, Tony Blair gave his first speech which mentioned CCS. He said he wanted "to advance work on promoting the development and uptake of cleaner energy technologies begun under the French presidency in 2003 and continued by the US this year". Blair said there was a need to invest heavily in existing technologies and also to stimulate innovation in new low-carbon technologies. He argued that there was "huge scope for improving energy efficiency and promoting the uptake of existing low-carbon technologies like PV, fuel cells and carbon sequestration" (Blair, 2004).

In March 2005, to aid conversations and build momentum, an international scientific conference was held at the University of Exeter titled "Avoiding Dangerous Climate Change". One topic covered was CCS (Gibbins et al., 2006).

The Chancellor of the Exchequer, Gordon Brown gave his first speech on the environment at the same time, and his pre-budget speech mentioned carbon capture as well, saying "British businesses can also be world leaders in environmentally-friendly technologies, such as in carbon capture and storage" (Brown, 2005).

In April, the new EU commissioner for energy, Andris Piebalgs, was advocating clean coal technology and CCS. As environmental journalist Fred Pearce noted, these were "now the EU's two top priorities in energy research, something that will anger environmentalists who want the world to abandon fossil fuels as quickly as possible" (Pearce, 2005).

In late 2005, a Clean Coal Task Group – a joint industry/unions/government advisory body – was formed. It was an initiative of a joint Trades Union Congress (TUC) and the Department for the Environment, Food and Rural Affairs (DEFRA) committee. Its remit was to "identify an appropriate policy framework and supporting economic instruments and regulatory framework that would take forward the research, development and promotion and initiation of clean coal burn and carbon capture and storage technologies".

In June 2006, it would release a report titled *A Framework for Clean Coal in Britain* (Clean Coal Task Group, 2006), adding more flow to the policy stream.

This period showed there was a broad (albeit cautious) belief that CCS is a "part of the solution"; it then becomes a question of – who is going to pay, how? This has remained the problem ever since.

4.3   Miller Field (aka "DF1")

In April 2005 an article appeared in the *Observer* newspaper extolling the virtues of Miller Field – a North Sea oil and gas area that was rapidly depleting – as a possible storage facility for captured carbon dioxide (McKie, 2005). A geologist was quoted as saying that since production at the Miller Field was coming to an end, *"we have a wonderful opportunity to develop techniques that could control global warming"*.

Despite this, when BP and Scottish and Southern Electricity (SSE) made a proposal for a CCS project, which would replace an existing gas-fired plant with a hydrogen plant, the announcement "appeared to take the UK government by surprise" (Haszeldine, 2012).

The proposed project was to capture CO_2 at SSE's power station in Peterhead and pipe it to Miller Field, 240 km offshore, using a pre-existing pipeline from the field (i.e. the direction of the flow of gas would be reversed). Power would be generated in a new 350 MV hydrogen-fuelled generator.

BP was enthused. Its chief executive, Lord Browne, said: "The project will offer a new, large-scale source of decarbonised electricity to consumers as well as extending the commercial life and contribution of the North Sea to the UK and Scottish economies" (Catan & Harvey, 2005).

It was called DF1 because it was expected that there would be a total of ten such projects (Shackley & Evar, 2012: 162).

Speaking to the UK Science and Technology Select Committee the following years, BP executive, Gardiner Hill, explained that Miller was well suited for the first large-scale industrial demonstration of offshore Enhanced Oil Recovery (EOR) since "CO_2 [was] indigenous in the oil, and so the platform and the production facility were built with CO_2 in mind" (Science and Technology Select Committee, 2006).

There were other hoped-for benefits, such as revisions to the new EU Emissions Trading Scheme to allow low-carbon schemes to trade their reduction alongside so-called zero carbon projects (Smedley, 2005). (This was to be an ongoing theme for CCS projects – attempting to tap into existing revenue streams such as the Clean Development Mechanism.)

This project was not the only one being mooted at the time. There were also proposals for projects in Lincolnshire (E.ON) Tilbury (RWE) and Teesside (Progressive Energy), but none of these had specific sites for the storage of captured carbon. Initially, all seemed well. In December 2005, Gordon Brown offered in-principle support (Anon, 2005). In early 2006, the Select Committee heard from the geologist Stuart Haszeldine as to the financial advantages that would accrue:

> Using Miller as a pilot could then unlock tax income for the Treasury of $1,400M from Miller over 20 years, and perhaps 5 times that from adjacent oilfields. If EOR is not undertaken as a by-product of CCS, then this income will never exist. In summary, if the consumer pays the cost of CO_2, the Treasury receives a type of stealth tax from the EOR, the UK gains in employment, and in diversity and security of supply.
> (Science and Technology Select Committee, 2006)

However, in July 2006, it emerged that two of the companies involved – Shell and Conoco-Philips – had pulled out (Forsyth, 2006). The following month BP warned it could abandon the Peterhead proposal if the government did not explain what incentives would be in place (Anon, 2006). The threat did not work – Chancellor Gordon Brown failed to give anticipated details of financial support for the scheme in his pre-budget report in December 2006 (Forsyth, 2007).

Eventually, having reportedly spent £50 million, BP announced it was abandoning the project. The final straw was the announcement in the 2007 Energy White Paper that details of a competition for funding of a CCS project would not be announced until November of that year. BP said that it had worked closely with DTI and had "put back its plans a number of times to try to meet the government's timetable. The decision to postpone the launch of a competition until November was a 'delay too far'" (Macalister, 2007).

There was a furious response from the Scottish National Party leader Alex Salmond, who said, "Never has so great an opportunity been passed up because of delay and incompetence and the inability of Westminster ministers to take decisions" (Wilson, 2007). Other commentators similarly bemoaned the government's decision-making (Clover, 2007; McKie, 2007). Stuart Haszeldine said, "The Government says it wants to lead a carbon capture and storage project. Well, it had a lead. Now it has a lead in hot air" (Prosser, 2007).

But the story is a little more complicated than selfless oil-company versus dithering government, and Labour did not accept the charge lying down.

Speaking on a Radio Scotland programme, former Energy Minister Brian Wilson said that he thought BP had been "quite clever about this because they have got out from under the opprobrium by and large about the decision. But it is a commercial decision by BP and BP has essentially decided to do it in Australia rather than here" (Dinwoodie, 2007).

Meanwhile, Alastair Darling, then Secretary of State for the DTI, said, "I just can't hand over a contract to one company. I would have liked to have seen this work done in Scotland because it would have been good for Scotland". He said that BP "have known for some time that it would be next year before we could announce a result". Darling pointed out that the only way he could have met BP's target was to have given them the contract, and if he had done that, he "would have been open to challenge from all the others" (Dinwoodie, 2007).

The CCSA remained relatively emollient – Jeff Chapman, chief executive of the Carbon Capture & Storage Association, said:

> It is sad but fortunately there are a lot of other schemes. We are always disappointed when the timetable is not as fast as we would like, but we accept that the government has a lot of work to do putting in place the appropriate regulations.
>
> (Macalister, 2007)

There is a further perspective here. Writing in 2022, two well-informed academics who had been involved in CCS research at the time and since speculated that it seems

> at least possible that if BP had followed advice to use the cheaper [post-combustion] option on the already-existing natural gas power plant at Peterhead then the project would have attracted the necessary government support – and UK CCUS deployment would be about 15 years ahead.
>
> (Gibbins & Lucquiaud, 2022)

Before turning to a discussion of the first CCS competition, which ran from November 2007 before petering out in 2011, a brief recap of the events in 2005 and 2007 is in order. Internationally, there was a huge upsurge in interest on climate change issues. In 2005, the EU's Emissions Trading Scheme had begun, and – with Russian ratification – the Kyoto Protocol finally had enough signatories to be law. The UNFCCC negotiating process could therefore begin to focus on what would happen next.

In September 2005, the IPCC's Special Report on CCS had been released, aiding proponents of the technology in making their case. In 2006, Al Gore's film *An Inconvenient Truth* was released, informing people who had been too young or too distracted during the 1988–1992 wave of concern about what climate change would mean. Finally, the IPCC's Fourth Assessment Report – the most emphatic to date – was released.

In the UK there were four events of note. First, after the May 2005 election, in which the Labour government had lost two-thirds of its majority, David Cameron became Conservative Party leader and chose to use environmental concerns as one way to "detoxify" the image of his party. Beyond cycling to work and a photo-opportunity in the Arctic, this also included support for CCS. In December 2007, Cameron gave a speech about clean coal in Beijing "developing green coal will be a priority for a Conservative Government: we will do what it takes to make Britain a world leader in this crucial field" (BBC, 2007a). Cameron's "vote blue go green" strategy led to a period of "competitive consensus in environmental decision-making, which will be discussed in the next section.

Second, in October 2005, the CCSA was created. The announcement was made during a two-day conference in London between international business leaders and senior UK ministers, organised by the British trade and environmental agencies. The 11 founding companies were British Petroleum, Progressive Energy, Air Products, Alstom Power UK, AMEC, ConocoPhillips, Mitsui Babcock, Schlumberger Oilfield UK, Scottish & Southern Energy, Shell Oil Co. and E.ON UK (Najor, 2005). The CCSA membership grew, and, as we shall see in this chapter and the next, it has played a crucial role in discussions, especially in the "picking up the pieces" years of 2015 to 2018.

Select committees continued to hold hearings into the technology, often simultaneously, examining different aspects of it. Worthy of note is the House of Science and Technology Committee, 2006. It heard from a range of experts, including Dr Jon Gibbins, the head of the UKCCS Consortium, an academic research group. He told MPs:

> The UK has the opportunity to make this technology acceptable possibly ten years earlier [than would otherwise be the case] and that could have huge implications when the globe is going to say, "Okay. It does not look too bad, tackling climate change; let's go for it".

Alongside such philanthropic motives, he also pointed to potential financial advantages, observing that

> there is an awful lot of money going to be traded. There will have to be projects to verify and a lot of financing for projects. A lot of that is likely to come out of the City of London. If we can get that experience here first, we can make some money for the UK.
>
> (Science and Technology Committee, 2006)

In late 2006, a report on the economics of climate change, commissioned by the Treasury, was released. Known as the Stern Review, after its author, World Bank economist Nicholas Stern, it argued that early action would be cost-effective. Finally, civil society action began to take place in the UK. An umbrella group, "Stop Climate Chaos", was launched, with membership of NGOs large and small. Beyond that, non-violent direct action protests began, with activists from the Camp for Climate Action focussing their attention on Drax coal-fired power station in Yorkshire. CCS was beginning to attract attention from leading environmental thinkers. In his 2006 book *Heat*, George Monbiot wrote:

> I have come to believe that this technology, alongside others which have been judged "too far away", can, with sufficient political commitment, be widely deployed long before 2030. The difficulties I have encountered while investigating the other technologies have persuaded me that carbon capture and storage – while it cannot provide the whole answer – can be and must be one of the means we use to make low-carbon electricity.

During this period, actors as diverse as coal miners unions, European Commissioners, international oil companies and environmentalists were advocating the technology. The sense of possibility, of necessity, is summed up well in Figure 4.1, from a presentation by Jon Gibbins and Hannah Chalmers at a seminar in September 2007.

Gibbins and Chalmers (2007) were not making predictions. Already by September 2007, with the experience of the Miller Field project, and the difficulties in the United States with the FutureGen project, it was clear things would not go smoothly. They were presenting a best-case scenario, with learning from demonstration plants cascading onwards. It was not, as we shall see, that CCS would follow its best-case path.

When this period began, the atmospheric concentration of carbon dioxide was 375.6 ppm, with an estimated annual amount of carbon dioxide being emitted by humans at 22.76 billion tonnes. At the end of the period the concentration was 363.5 ppm, with emissions at 24.3 billion tonnes.

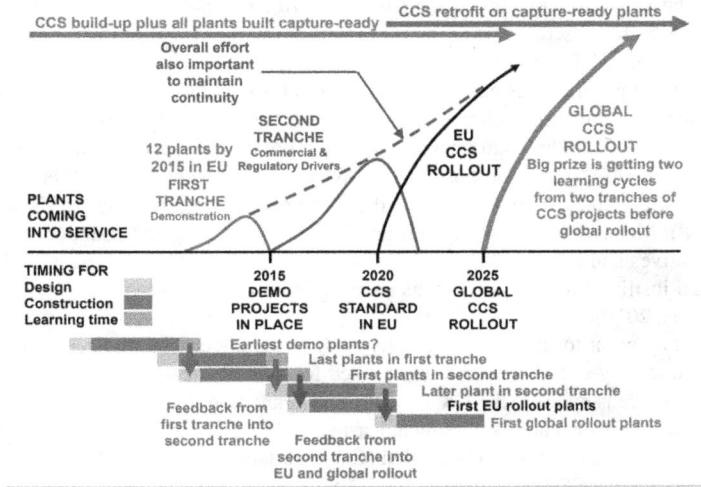

Figure 4.1 The future of CCS

Source: From Gibbins and Chalmers (2007), with the kind permission of Jon Gibbins.

4.4   The first competition: 2007–2011

The period now discussed is one of intense activity – and, ultimately, intense disappointment both internationally and nationally.

In January 2007, the European Commission proposed that by 2015 there would be 12 demonstration plants in operation (Pearce, 2008). These would provide useful learning – at technological, economic and policy levels – for a much larger roll-out to then occur. The story of this policymaking process, and its aftermath (no plants were built), is told well by Boasson and Wettestad (2014). It includes the important point that not all those pushing for CCS were particularly enamoured of the technology. It quotes prominent Member of the European Parliament Chris Davies, then pushing for CCS to be supported by the EU, as saying:

> I hate CCS. . . . It is just that I hate coal more. We have to promote CCS. China, India and the US need to realise that they will have to pay a lot more if they want to use coal.

The exuberance of the time is exemplified by the fact that Richard Branson created the US$25 million Virgin Earth Challenge – launched in London with Al Gore in February 2007 – to find commercial solutions for extracting greenhouse gases out of the air (BBC, 2007b). Ultimately, no prize money was ever dispensed.

Climate policy had become headline news in the UK, with new reports, television programmes and initiatives – from business, civil society and governments – announced almost daily. It was, in the terms of Kingdon, the case that a policy window had opened. The main policy advanced in this window was the 2008 CCA. The push for an Act was broad-based, with Friends of the Earth leading a coalition of civil society organisations (Carter & Childs, 2018). Meanwhile, all three main political parties in England (Labour, Conservatives and Liberal Democrats) were agreed on broad principles, with the usual jostling for credit. This was an era of competitive consensus (Carter & Jacobs, 2010).

They point to just how much changed; in the 2006 Climate Change Programme only £35 million was earmarked for development of CCS technologies, whereas by 2009 the UK Low Carbon Transition Plan promised no new coal-fired power stations without at least partial CCS and up to 4 CCS demonstration plants to be publicly funded (Carter & Jacobs, 2010: 130).

When the CCA was passed in 2008, it increased the "60% emissions reductions by 2050" target that had been announced in the 2003 Energy White Paper to 80% (see Chapter 5 for a discussion of what moving from an 80% target to "net zero" by the same date meant for industry attention to the agenda). It also created an independent CCC, which was to report to parliament on issues related to achieving the new target and what progress was being made on achieving the five-yearly carbon budgets that the Act instituted. The very first report of the CCC (CCC, 2008) stated that CCS was a highly important technology, and it has not resiled from that line.

Analyst Chris Littlecott argues that the CCC's recommendation in Chapter 5 in that first report that the UK power sector should be "almost entirely decarbonised by 2030" was "of singular importance for CCS" and "provided NGOs with a powerful supporting argument for their campaign against new unabated coal, while also requiring that they gave consideration to the potential mix of low-carbon technologies that could be delivered by 2030" (Littlecott, 2012: 427).[2]

As Inderberg and Wettestad (2015) note, the CCA also prepared the ground for Energy Market Reform (EMR), which included a "Contracts for Difference" (CfD) scheme,

an adjustable feed-in premium system scheduled to start in 2017. Supported technologies (including CCS) will sell electricity and heat in the market, and CfD will cover the difference between the estimated market price and the long-term price needed to promote investments in a given technology – the "Strike Price".

The CfD mechanism became a crucial part of the bid to create new funding arrangements for CCS, as will be discussed in Chapter 5 (and the future prospects will be discussed in Chapter 6).

Just before the CCA gained Royal Assent, there was a significant rearrangement of Departments of State. A new Department of Energy and Climate Change (DECC) was created in October 2008 and "resulted in attempts to move UK CCS policy into a more proactive role as part of UK decarbonisation strategy in line with carbon budgets", noting meanwhile that CCS policy remained almost entirely focussed on coal CCS for power generation, rather than including gas and industrial CCS (Littlecott, 2012: 432).

In April 2009, Secretary of State for DECC, Ed Miliband, announced plans for a CCS levy to fund the development of various projects.

By 2008, there was growing civil society antipathy towards coal. One way that the government tried to manage this concern was by stating that the new power stations could only be built as long as they were "capture-ready". The formulation did not satisfy many, and there was spirited public resistance at Kingsnorth power station in August 2008, alongside stunts by Greenpeace – with six of its activists acquitted in late 2008, following an October 2007 action (Vidal, 2008).[3]

In essence, the first CCS Commercialisation Competition was a prolonged version of the bloody mystery novel *And Then There Were None* by Agatha Christie (1939). In that novel, eight guests and two staff on a mysterious island were killed off, one by one, trying to figure out who their assailant was. The competition, meanwhile, attracted nine entrants. This was then whittled down to four. Each of these, over three tortuous years, dropped out, until there were none.

This section will cover the rationale for the competition, its announcement by Prime Minister Gordon Brown in November 2007, the closure of the application process at the end of March 2008 and then the gradual dropping away of entrants. It will also touch on the coming of the coalition government in May 2010. It closes with explanations for the failure of the process.

The staging of a competition is one way of keeping ministers at "armslength" from awarding possibly very lucrative contracts, thus insulating them from accusations of picking winners (other motives will be discussed later). Even before the competition formally began, it was "mired with disappointment within the industry". There were companies developing pre-combustion technologies frustrated at the narrowing of the competition to focus on post-combustion only. In October 2007, a group of energy companies, led by Centrica, had said they would challenge the government's decision (Webb, 2007).

The competition launch announcement was made by Gordon Brown at an event hosted by the World Wildlife Fund on 19 November 2007. The expectation was that the winner would be appointed in summer 2009 to begin developing the CCS demonstration. The competition was restricted to 300 MW post-combustion carbon capture projects funded by coal. The rationale was that post-combustion carbon capture could be retrofitted to hundreds of existing plants throughout the world. This was especially important given that so many new coal plants were being constructed in India and China.

John Hutton, Secretary of State of Business and Enterprise, was quoted as saying:

> Our analysis shows that post-combustion capture is the most relevant technology to the vast proportion of coal-fired generation capacity globally. A commercial-scale demonstration of this technology, as part of a full CCS chain, opens up huge possibilities, not just for Britain but also for the world.
>
> (Adam, 2008)

Similarly, John Gale of the IEAGHG, which had started the ball rolling in the early 1990s, concurred, saying, *"The UK sees a massive market overseas for this. China and India are putting plants in that could have to be retrofitted at some stage in the future. Someone is going to be able to sell that technology"* (Adam, 2008).

Applications to take part in the contest closed on 31 March 2008. There were nine projects entered, and according to Haszeldine (2012), the government seemed "somewhat shocked by the seriousness and number of industrial interested parties". The process was, tongue-in-cheek, likened to a reality TV show.

> The crowning of the competition's eventual winner is unlikely to make good television, but it could be make or break time for the planet. Call it the ultimate reality show. Although Britain's claims that the competition will produce the first plant of its kind will probably be overshadowed by progress elsewhere in coming years, the plan remains one of the first serious attempts to introduce large-scale carbon capture and storage.
>
> (Adam, 2007)

Criticism of the competition criteria centred on the lack of breadth and the lack of ambition. As the centre-right think tank *Policy Exchange* pointed out, the cautious wording of the competition rules means that the winning project need only have 50 MW of power plant operating with CCS by the end of 2014, with the full capacity only "as soon as possible thereafter" (Adam, 2008).

Not everyone was convinced this was the best framing. In July 2008, before the shortlist was announced, MPs on the Environmental Audit Committee had said the "decision to limit funding to a single project has led to the loss of a number of promising projects developing other forms of technology" – and they accused ministers of dragging their heels. SNP's business spokesman Mike Weir said that the government was obsessed with developing a technological solution for coal-fired power stations in China, ignoring the need to clean up gas-fired power stations at home (Perry, 2008; see also Hughes, 2012).

The Conservative Party was scathing. In a Green Paper it pointed out that the House of Commons Science and Technology Select Committee had concluded the following in 2006:

> Multiple full scale demonstration projects using different types of capture technology and storage conditions are urgently needed. . . . Picking one winner for a small-scale demonstration project reveals a profound lack of ambition and fails to create an adequate framework for the transition to low carbon generation. Clearly, a more effective framework for developing our CCS potential is required.
>
> (Conservative Party, 2009)

The Conservative answer, invoking the Californian policy of Governor Arnold Schwarzenegger, was to propose something broader. Opposition leader David Cameron said he would replace the current government's plans for a single-plant CCS competition and instead "fund a variety of large-scale CCS demonstration projects of varying types of technology, including both pre- and post-combustion carbon capture over the next five to 10 years, out of receipts from the third and subsequent phases of EU Emissions Trading Scheme" (Conservative Party, 2009). The contrast between the enthusiasm in opposition and the performance in government is something that shall be covered later in this chapter and again in Chapter 5.

After the 31 March deadline, the various bids were considered. At the beginning of July the shortlist of four was announced. The four selected entrants were BP Alternative Energy, E.On UK, Scottish Power Generation and Peel Power. BP Alternative Energy planned to enter a 475 MW gas-fired plant at Peterhead. E.On entered its planned clean-coal power plant upgrade project at Kingsnorth, near London. The company was intending to construct two new 800 MW clean-coal units, both "capture-ready". A consortium led by Scottish Power planned to develop a CCS-ready coal-fired power station, probably at Longannet in Scotland, while Peel Power proposed a coal-fired power station (Anon, 2008a).

The process was beset by slippage of deadlines and uncertainty over funding sources. In addition, unsuccessful candidates, such as RWE Npower, sought judicial review (Wood, 2008).

Writing in late 2008, Bowman and Addison (2008: 521) expected the decision of DECC as to the winner of the demonstration competition in "summer of 2009". It was not to be.

BP pulled out in December 2008, 17 months after withdrawing its Miller Field project. It said that it could not find the "right mix of players to create a winning consortium" Harvey, 2008. The newly created DECC put a brave face on matters, saying that BP's decision did "not compromise the integrity of the competition nor will it have a material impact on the robustness of the procurement process. It plans to continue the competition with the remaining participants" (Anon, 2008b).[4]

In April 2009, following advice from the CCC, the government changed the competition rules to allow at least two demonstrations, including one using pre-combustion. However, according to Littlecott (2012: 433), the launch of the call for projects was repeatedly delayed due to uncertainties over funding availability, with consequences for being able to access EU funding.

In November 2009, during the Copenhagen Climate Summit, Peel pulled out. This left only two bidders, E.On and the now Spanish-owned Scottish Power project. In March 2010, funding was awarded to the two remaining entrants in the competition for front-end engineering and design (FEED) studies on the Kingsnorth and Longannet projects, respectively.[5]

It is important to remember the broader context at this point. The Copenhagen Climate Conference, at which a successor agreement to the Kyoto Protocol was supposed to be agreed, had ended in acrimony. Only the most vague of documents, the Copenhagen Accord, emerged. It was essentially no more than an agreement to keep on having meetings. Investors and governments, also dealing with all the uncertainties of the Global Financial Crisis and its aftermath, were becoming less enthused about climate and CCS. As the *Financial Times* noted, "many countries over the past year [have] moved away from curbing greenhouse gas emissions towards a focus on economic growth and jobs" (Crooks & Pfeifer, 2011).

At this point, it was still believed that "a process to select a further three demonstration projects is due to begin later this year and finish in 2011" (Smith, 2010: 101).

The change in government after the 2010 general election and the coming of a coalition government made up of the Conservative Party and the Liberal Democrats did not create much uncertainty (though the Comprehensive Spending Review, later that year, would). This was in part due to the Liberal Democrats being keen on climate action and the presence of an energy minister, the Conservative Charles Hendry, who had been closely following the issue while in opposition.

Another minister, Gregory Barker, gave a robust and optimistic statement of sustained ambition, saying in the House of Commons that

> the coalition Government are committed to carbon capture and storage, which will be a major plank in our efforts to decarbonise our energy supply by 2030; we are committed to the generation of 5 GW of CCS by 2020. We see the potential of CCS, not just for our domestic use and as part of our plan to decarbonise the economy, but as a huge potential export industry for the UK in which we can not only capture new markets for British jobs, but help the world in striving to decarbonise the global economy.

Such a rousing statement did not prevent E.On, in October 2010, from withdrawing from the competition. E.On chief executive Paul Golby said:

> Having postponed Kingsnorth last year, it has become clear that the economic conditions are still not right for us to progress the project and so, simply put, we have no power station on which to build a CCS demonstration. We therefore took the decision to withdraw from the government's competition because we cannot proceed within the competition timescales.
>
> (Stone, 2010)

This left only the Longannet project of Iberdrola. Finally, in October 2011, to no-one's particular surprise, they declared that they too were withdrawing their project from the competition. It was for a variety of reasons, including costs and also who should be responsible for various risks in long-term storage that were not yet well-defined (Kemp, 2012). Others pointed to the fact that Iberdrola was much less committed to coal generation and more focussed on hydro-power, wind and gas combustion. The March 2011 abolition of a proposed CCS levy meant that the potential funding streams were becoming steadily more opaque and uncertain. It had been introduced by the previous Labour government in its March 2010 Energy Act (see Gibbins & Lucquiaud, 2022 for discussion of this), and its abolition is another sign of the power of the Treasury in policymaking around CCS, albeit an unsurprising one, given the Treasury's traditional antipathy towards ring-fencing of funding for specific purposes.

4.4.1 Why did the competition fail?

It is easy, with the benefit of a decade and a half's hindsight, to blame policymakers for indecision and incompetence. It is easy to forget that policymaking involves many competing priorities, with a lack of information and external factors (in this case the Global Financial Crisis and the failure of the Copenhagen meeting) complicating matters. As Littlecott (2012: 433) notes, both DECC and its Office for Carbon Capture and Storage (OCCS) were "repeatedly forced to readjust to changed financial circumstances".

More specifically, though, there were other issues, Littlecott (2012: 431) argues, that "CCS was treated by officials more like a risky R&D project than an inescapable component of a decarbonised energy system . . . combined with an historical preference for 'letting the market decide', this resulted in insufficient investor confidence in the policies put forward".

Hughes (2012) argued that the focus on CCS as a source of exportable intellectual property and skills rather than as a green technology was the root cause of the policy mistakes. In March 2012, the parliamentary watchdog organisation the National Audit Office (NAO) cited insufficient planning and lack of financial clarity and recognition of commercial risks (NAO, 2012; Inderberg & Wettestad, 2015).

Even before the competition petered out, there had been more sceptical perspectives on what was unfolding. As Scrase & Watson, 2009: 176) state, "It is possible that the way the competition was handled was in part a deliberate tactic to postpone spending the large sums of money involved". They even suggest that the complicated and restrictive design process might even be a signal to potential nuclear power investors that CCS would not be a threat to their ability to sell "low-carbon" electricity to consumers. This theme was taken up in 2012 by Stuart Haszeldine, who observed that

> the design requirements for the power plant and CCS operation, and the financial restrictions were all extremely tightly written, with little room for innovation or development. At higher levels of government, especially in the Treasury, it must have been well known at the time that these types of specification are extremely hard to meet. Is it possible, then, that a deliberately difficult hurdle for CCS power plant to surpass was set, in full knowledge that such a project would almost inevitably fail?
>
> (Haszeldine, 2012; emphasis added)

Haszeldine notes that at this time there was also "a strong and active discussion to form the direction for construction of new nuclear power plant within the UK", which many in the government saw as a source of cheap, reliable low-carbon electricity. In the absence of an unprecedented leak or confession, it is the case that, as Haszeldine drily notes, "the true rationale behind these decisions may never be fully known" (Haszeldine, 2012).

If the causes of the failure of the competition are unclear, or in dispute, the consequences were not. There was a lost opportunity to build not only supply chains but also skills and expertise for a much larger roll-out (recall the figure from Gibbins and Chalmers, earlier in this chapter). Interviewed for a research project about CCS, one informed participant told the researcher, "If you did two 300 MW projects by 2014 and another two by 300 MW for 2019 and that was it, then no, you wouldn't have the human capital ready to do it, for full roll out by the 2020s" (Evar, 2011: 3421). The participant said that, to facilitate the necessary learning, as many as 15–20 demonstration plants would be necessary in the UK. The reality was that there would be no projects by 2019, let alone 15–20.

Given the obvious inadequacy of the first competition, and amid plans for more demonstration projects, a second competition was being planned, continually announced and then pushed back. To try to expedite this, even before the Longannet project ended, the CCSA sent a letter to the UK's Minister for

Energy and Climate Change, Chris Huhne, in June 2011. The CCSA's Chief Executive Jeff Chapman noted that

> the CCSA Board is concerned that industry does not yet see firm Government signals about the long-term viability of a CCS market, particularly when compared to the signals given for nuclear and renewables. . . . Without clear recognition from Government that CCS will be part of the generation mix, we are concerned that industry's appetite for continued commitment will shrink. We would urge you to say that roll out (of CCS projects beyond the 4 demonstration projects planned for the UK) should be planned on the basis that Projects 1–4 will be successful and will yield valuable learning through planning, design, build and operation.
>
> (Chapman, 2011)

Three months after this letter, the CCSA would release a document, *A Strategy for CCS in the UK and Beyond*, with the objective of pushing for the creation of infrastructure that could capture 500 Mt per year by 2030. This could, the CCSS argued, create a market worth £10 billion/year for UK plc by 2025, with more than 50,000 quality jobs by 2030.

At this stage, the government was supposed to be releasing its own roadmap on 17 November (Bellona, 2011). It would not be released for another five months, until early April 2012.

In early 2012, the Deputy Director of the CCSA, Luke Warren, was looking to the future, beyond the CCS as a technology only for abating coal and gas power production.

> [T]he provision of CO_2 infrastructure – resulting in several regional pipeline hubs servicing multiple clusters of emission sources and delivering the CO_2 to a network of stores – will be the key to the widespread and cost-effective adoption of CCS in the UK. Policies that support the development of appropriate CO_2 infrastructure as part of the first capture projects must be considered a national priority; we are delighted that the Government now explicitly recognizes the value of clusters in the CCS program. These first projects will provide a one-off opportunity to cost-effectively create reliable CO_2 transport and storage infrastructure. Finally we must not forget that power plants are only one source of UK CO_2 emissions; there are also very significant CO_2 emissions from the steel, cement, refining, and chemicals industries. For a large proportion of these industries, there is no realistic means of decarbonization other than CCS, because the CO_2 is emitted from the process as well as the fuel consumed. Furthermore, in many cases, the CO_2 sources have relatively high concentrations of CO_2, resulting in relatively lower capture costs. At the Industry Day the Government recognized the importance of these sources of CO_2 and the opportunity for them to be integrated into regional clusters.
>
> (Warren, 2012: 4)

King Coal's quixotic attempt to introduce clean coal

Amidst the multinational companies proposing projects and entering the various competitions, mention should also be made of Richard Budge. An entrepreneur, he had bought up 17 deep coal mines when they were sold off in the early 1990s, earning the nickname "King Coal". Through the 1990s he was advocating for (and seeking government funding for) clean coal technologies. After leaving RJB Mining in 2001 he bought Hatfield Colliery, with the aim of building the UK's first clean coal power station. He was not successful in this first effort but restarted the colliery in 2006 under the name of Powerfuel. Russian coal producer Kuzbassrazrezugol bought a 51% stake in Powerfuel, and the following year production restarted at the mine. Budge again was seeking government funding, saying, "New power station development of this type, without a change in Government policy, will be the exception" (Prosser, 2007). A blow to the project was its ineligibility for the first CCS competition because it was planning pre-combustion rather than post-combustion technology. Powerfuel was the only UK CCS project to get European funding, but it was unable to get the rest of the money it needed and went into administration (Mason, 2010). The assets were bought by another company, 2CO, but it too failed to secure the required government support, and the CCS project foundered. Budge died in 2016.

4.5   Second competition: 2012–2015

On 3 April 2012, DECC released its long-awaited (and frequently delayed) CCS roadmap. The 50-page document included 16 mentions of clusters (something that would later come to dominate CCS policy). DECC said they were "seeking to support the development of a sustainable CCS industry that will capture emissions from clusters of power and industrial plants linked together by a pipeline network transporting CO_2 to suitable storage sites offshore" (DECC, 2012a: 6). The hope was to see the first demonstration plants come onstream between 2016 and 2020, with the equivalent of 12–20 large power stations fitted with the technology by the end of the 2020s.

DECC also released details of the second CCS competition. The same £1 billion would be available to support the upfront costs of early projects, along with a commitment to further funding through low-carbon Contracts for Difference. Alongside this, there was the £13 million UKCCS Research Centre (UKCCSRC), which has proved to be an important part of the academic ecosystem and a crucial player in the difficult days of 2016, discussed in the next chapter. The reaction was largely positive. The Scottish Energy Minister Fergus Ewing welcomed it while making sure to mention "two previous false starts" (Anon, 2012). The CCSA was also broadly supportive, noting that the Roadmap recognised both the importance of transport and storage infrastructure and also the importance of decarbonising key industrial sectors. Academic Jim Watson was cautious – noting that "we still don't know when these technologies will be technically proven at full scale, and whether their costs will be competitive with other low-carbon options". He found it

> encouraging to see DECC revising the competition to develop commercial scale CCS technologies and, even more importantly, expanding the entry criteria to include technologies that rely on gas and coal – apparently in recognition of the importance of keeping options open, and not closing down on any particular variant at such an early stage in the process,

Dustin Benton of Green Alliance was also supportive and pointed to the creation of a cost reduction taskforce and greater clarity about long-term financing steps in the right direction but cautioned that "*positive language about the potential role of clusters will need to be matched by action*".

By this time, Richard Budge's Powerfuel had been taken over by 2CO Energy. It argued that the government needed to take a "cluster approach" to supporting CCS projects in the UK as economies of scale would allow the UK's clean electricity goals to be met at least cost to the UK. It said that it had "already held discussions with other project developers to agree a shared vision for a major CCS cluster in the Humber Gateway region linking together multiple power plants and industrial emitters via a shared pipeline infrastructure to transport captured CO_2 offshore to North Sea storage sites".

All these above quotes are from Hickman (2012). In closing the blog Hickman pointed to the danger that

> CCS will be used to help recover "hard to reach" oil and gas from the North Sea. Will this simply cancel out any emissions reductions achieved by capturing the CO_2 emitted from power stations? And how much energy will be used – and efficiency lost – capturing, pumping and storing carbon emissions from power plants?

A month later, it was announced that the CCSA's Jeff Chapman would chair the *Cost Reduction Taskforce* that had been announced as part of the CCS Roadmap. The taskforce's role was – as per its name – to help the government reduce costs. The taskforce released its report the following year, and all seemed well. (A successor to the *Cost Reduction Taskforce* – the *Cost*

Challenge Taskforce – would play a crucial role in helping revive CCS a few years later, as we shall see in Chapter 5.)

Eight bids were submitted by the competition's deadline of 3 July. Four were selected "after a thorough evaluation process that considered project deliverability, value for money, and the Government's timetable to deliver a cost-competitive CCS industry in the 2020s" (DECC, 2012b). The four were as follows:

> The Captain Clean Energy project, a proposal for a 570 megawatt power plant converting coal into gas in Grangemouth, Scotland; Teesside Low Carbon project, another coal gasification project, with carbon emissions stored in a depleted oil field and a saltwater aquifer, led by Progressive Energy and involving GDF Suez, Premier Oil and BOC; Shell and SSE's plant to retro-fit a 340 megawatt carbon capture scheme as part of an existing 1180 megawatt gas power station at Peterhead, Scotland and finally the White Rose project at a proposed new 304 megawatt coal-fired power station, led by Alstom and involving Drax, BOC and National Grid.
>
> (Beament, 2012)

The government stated that the competition would be completed the following year (Barrett, 2012). This was not to be the case.

Meanwhile, Labour claimed that the money would not be forthcoming and that the Treasury's refusal to confirm co-funding had cost those short-listed projects which had applied for European funding the opportunity to access the 600 million euros in matched funding. In December 2012, Shadow Energy and Climate Change Minister Tom Greatrex cited a previously unpublished Cabinet Office project assessment review he had obtained as saying only £200 million was available, not the £1 billion. He asked, rhetorically, "how does the minister expect there will ever be progress in developing commercial CCS if the Government's financial commitment falls so far short of the Prime Minister's warm words?"

When Greatrex repeated the question in June 2013, in the context of an imminent spending review, Energy and Climate Change Minister Greg Barker, who had made such bold predictions of CCS's future in 2010, rejected this as baseless scaremongering (Sculthorpe, 2013).

In between these efforts at highlighting the possible non-appearance of the funding, in March 2013, the UK government had announced the two preferred bidders, Peterhead gas CCS project and the White Rose oxyfuel project. At that time, the expectation was that "*In early 2015, it is hoped there will be FIDs on the construction of up to two projects*" (Anon, 2013).

In October 2013, Members of the Scottish Parliament were pondering what Plan B might be if CCS did not arrive. Green MSP Patrick Harvie said that there had been times when the proponents of the technology had talked about it as if it were already present. "*They've talked about the incredibly important role it can play, not the incredibly important role it might play"*. He also warned that CCS must not be used as a pretext to justify increasing fossil-fuel energy-generating capacity now, before the technology is even available (McLaughlin & Bews, 2013).

The following month, one of those proponents, Stuart Haszeldine, was expressing exasperation at the slow pace of the second competition, saying:

> We're about a year behind on a one-year project where we have a unique resource and should be getting on with it about 10 times faster. Three-quarters of the Peterhead project has already been designed and inspected as part of previous schemes. . . . The whole delay is caused by the DECC process. . . . We have slipped from first place as a potential world leader in this technology to a very poor fourth.

Northeast MP and SNP energy spokesman Mike Weir joined in, adding that in his view "It does suggest that the hand of the Treasury is behind the delays with a reluctance to actually commit money promised" (Mckiernan, 2013).

In the same month, the TUC, which had been supporting CCS since 2005, combined with the CCSA to release *A UK Vision for Carbon Capture and Storage* (TUC/CCSA, 2013), trying to counter a trend towards decreasing policymaker appetite noted by the Global Carbon Capture and Storage Institute (Anon, 2013). The TUC/CCSA document was based on modelling of the economic benefits of CCS for the UK commissioned by the Energy Technologies Institute. It argued that without CCS, the cost of delivering a UK low-carbon energy mix in 2050 would increase by £30–40 billion per year (1% of GDP).

At this stage, a very important development began to emerge. Teesside's project had not been shortlisted. As a kind of consolation prize they were given £1 million, as part of a City Deal, to explore CCS technology. A spokesperson for the local enterprise partnership Tees Valley Unlimited (TVU) told a journalist:

> Energy intensive companies on Teesside did not want to lose an opportunity to develop a CCS infrastructure which would significantly reduce their carbon emissions. They came together to develop an alternative strategy to progress industrial CCS. With rising carbon prices, mechanisms such as CCS need to be in place so industrial companies can reduce their carbon emissions without reducing production.
>
> (Price, 2014)

The Teesside work, under the banner of the "Teesside Collective", would prove to be crucial in keeping the CCS flame alive in the period 2016–2018.

The competition ground on FEED contracts was eventually signed in December 2013 and February 2014 respectively for White Rose (Capture

Power Limited) and Peterhead (Shell). White Rose received 300 million euros in June of 2014.

Alongside the competition, other activities around CCS and industrial decarbonisation were underway. In 2013, two government departments, DECC and Business Innovation and Skills (BIS), responding to calls for increased focus on reducing industrial emissions, had launched a process of consultation for eight sectors (including cement, glass, steel and chemicals) on how they might create industrial decarbonisation and energy efficiency roadmaps that would take them to very much reduced emissions from their production processes by 2050. The reports were released in March 2015, just before the general election, which ended the coalition and gave the Conservatives a majority in their own right.

It seemed that, at last, the rubber was hitting the road. Various think tanks and bodies were producing reports about what else needed to be done. Yes, there continued to be minor hiccups, such as Drax withdrawing from the White Rose project, but this was not seen as particularly consequential. The future was tolerably bright.

And then, on 25 November 2015, everything changed.

Notes

1 Speaking in 2022, a participant reminisced. "There was great enthusiasm for carbon capture technologies, particularly in the United States (Future-Gen), where a significant clean coal power plant project was envisioned. However, this program never came to fruition, and interest in CCS gradually declined" (Rustici, 2023).

2 Various CCC statements about the indispensability of CCS, like those in the IPCC's Special Report on CCS, often are used by advocates of the technology since they have the imprimatur of an independent and highly credible body.

3 For more on the role of the Green Alliance in bringing together industry and environmental NGOs to discuss the UK's decarbonisation dilemmas, see Littlecott (2012).

4 "The final BP withdrawal from any CCS project in the UK, was timed for the day of budget announcements on CCS support level and rate of progress. In a similar way E.ON chose to withdraw their Kingsnorth project on the day of the Comprehensive Spending Review announcements. Those timings suggest that these companies had already made their analyses, and were seeking high profile exits to pointedly make their case on the inadequate match between UK ambition and finance. Those warnings were ignored" (Haszeldine, 2012: 445).

5 The last days of the Brown government also saw the announcement of "low carbon economic areas" – a curious rebirth of regional and industrial policy and rhetoric which did not survive the "bonfire of the quangos" that ensued. The basic ideas, though, re-appear later as CCS clusters. As well as the venerable RCEP and the more recent Sustainable Development Commission

being abolished, Regional Development Agencies were axed. This had consequences for Abolition of the RDAs. In the eyes of one participant, reported in Gough and Mander (2022): "I think we had the regional Yorkshire and Humber cluster, and getting rid of the Regional Development Agencies was just an unmitigated disaster as far as CCS was concerned because there you had a cluster that would get together, the gel was the Regional Development Agency".

References

Adam, D. 2003. Plan to bury CO_2 under North sea. *The Guardian,* September 5. https://www.theguardian.com/science/2003/sep/05/sciencenews.science

Adam, D. 2007. The carbon question: Chance for Britain to lead the world: A competition to build Britain's first full-scale, clean coal power plant could have a global economic and environmental impact. *The Guardian,* June 25 (CCS supplement), p. 9.

Adam, D. 2008. The carbon question: Chance for Britain to lead the world: A competition to build Britain's first full-scale, clean coal power plant could have a global economic and environmental impact. *The Guardian,* June 25 (CCS supplement), p. 9.

Ali, T., et al. 2004. Letter: New Weapon against Global Warming. *The Independent,* May 27, p. 38.

Anon, 2004. UK considering offshore storage of CO_2 from generators, says minister. *European Daily Electricity Markets,* August 2.

Anon. 2005. Peterhead 'clean power' proposal gets backing. *Aberdeen Press and Journal,* December 6, p. 10.

Anon. 2006. Come off it with such threats, BP! *Aberdeen Press and Journal,* September 4, p. 18.

Anon. 2008a. UK picks four for CCS. *International Oil Daily,* July 2.

Anon. 2008b. BP gives up on UK's CCS effort. *Modern Power System,* December 23, p. 5.

Anon. 2012. Carbon capture and storage. *European Union News,* April 3.

Anon. 2013. Carbon capture and storage – CCS, not dead but running out of steam. *Modern Power Systems,* December 1, p. 12.

Barrett, E. 2012. Government Funding "Falling Short". *Press Association Mediapoint,* December 13.

BBC. 2007a. All coal-fired power stations would be fitted with "green coal" technology under a Conservative government, David Cameron has said on a trip to China. *BBC,* December 20. http://news.bbc.co.uk/1/mobile/uk_politics/7153139.stm

BBC. 2007b. Branson launches $25m climate bid. *BBC,* February 9. http://news.bbc.co.uk/1/hi/sci/tech/6345557.stm

Beament, E. 2012. Carbon capture schemes shortlisted. *Press Association Mediapoint,* October 30.

Bellona. 2011. New CCS Strategy for the UK launched by CCSA. *Bellona,* September 11. https://bellona.org/news/ccs/2011-09-new-ccs-strategy-for-the-uk-was-launched-by-ccsa

Blair, T. 2004. Full text: Blair's climate change speech. *The Guardian*, September 15. www.theguardian.com/politics/2004/sep/15/greenpolitics.uk

Boasson, L. and Wettestad, J. 2014. Policy invention and entrepreneurship: Bankrolling the burying of carbon in the EU. *Global Environmental Change*, 29, 404–412.

Bowman, J. and Addison, J. 2008. Carbon capture and storage – "the only hope for mankind?": an update. *Law and Financial Markets Review*, 2(6), 516–522. http://dx.doi.org/10.1080/17521440.2008.11428011

Brown, G. 2005. Full text: The chancellor's budget speech. *The Guardian*, March 16. www.theguardian.com/politics/2005/mar/16/economy.uk

Carter, N. and Childs, M. 2018. Friends of the Earth as a policy entrepreneur: 'TheBig Ask' campaign for a UK Climate Change Act. *Environmental Politics*, 27(6).

Carter, N. and Jacobs, M. 2014. Explaining radical policy change: The case of climate change and energy policy under the British labour Government 2006–2010. *Public Administration*, 92(1), 125–141.

Catan, T. and Harvey, F. 2005. Hydrogen power plant planned for Peterhead. *Financial Times*, July 1, p. 24.

CCC. 2008. Building a low-carbon economy – The UK's contribution to tackling climate change. *The first report of the Committee on Climate Change*, December 2008, www.theccc.org.uk/pdf/7980-TSO%20Book%20Chap%205.pdf

Chapman, J. 2011. Letter from the Carbon Capture and Storage Association, London, to the Right Honourable Chris Huhne, MP, Secretary of State for Energy and Climate Change, dated 29 June 2011. Cited in Shackley and Evar, (2012: 161).

Christie, A. 1939/2004. *And Then There Were None*. London: Macmillan.

Clean Coal Task Group. 2006. *A Framework for Clean Coal in Britain*. www.tuc.org.uk/research-analysis/reports/framework-clean-coal-britain

Clover, C. 2007. Earthlog Trouble ahead. *The Daily Telegraph*, May 25, p. 4.

Conservative Party. 2009. *Green Paper*. London: Conservative Party.

Crooks, E. and Pfeifer, S. 2011. Capture technology faces a more hostile environment. *Financial Times*, January 17 (FT REPORT – INNOVATION IN ENERGY), p. 3.

DECC. 2012a. *The CCS Roadmap*. www.gov.uk/government/publications/the-ccs-roadmap

DECC. 2012b. *Short List for UK's £1bn CCS Competition Announced*, October 30. www.gov.uk/government/news/short-list-for-uk-s-1bn-ccs-competition-announced

Dinwoodie, R. 2007. Wilson blames BP for carbon capture pull-out. *The Herald*, May 26, p. 6.

DTI. 2003. *Review of the Feasibility of Carbon Dioxide Capture and Storage in the UK*. London: HMSO. https://publications.parliament.uk/pa/cm200506/cmselect/cmsctech/578/578m.pdf

Evar, B. 2011. Conditional inevitability: Expert perceptions of carbon capture and storage uncertainties in the UK context. *Energy Policy*, 39(6), 3414–3424.

Forsyth, I. 2006. Green power plant still on despite firms pulling out. *Aberdeen Press and Journal*, July 1, p. 24.
Forsyth, I. 2007. Scdi power project plea to treasury. *Aberdeen Press and Journal*, January 19, p. 21.
Gibbins, J. and Chalmers, H. 2007. Carbon capture and storage. *Presentation at BIEE Energy White Paper Seminar*, September 25. British Institute of Energy Economics, https://www.biee.org/wp-content/uploads/Carbon_Capture_and_Storage-2007.pdf
Gibbins, J., Haszeldine, S., Holloway, S., Pearce, J., Oakey, J., Shackley, S. and Turley, C. 2006. Scope for future CO_2 emission reductions from electricity generation through the deployment of carbon capture and storage technologies. In *Voiding Dangerous Climate Change*, edited by Schnellnhuber, H. J. Cambridge: Cambridge University Press
Gibbins, J. and Lucquiaud, M. 2022. The development of UK CCUS strategy and current plans for large-scale deployment of this technology. *Annales des Mines – Responsabilité et Environnement*, 105(1), 26–30.
Gough, C. and Mander, S. 2022. CCS industrial clusters: Building a social license to operate. *International Journal of Greenhouse Gas Control*, 119.
Harvey, F. 2008. BP scraps carbon capture project. *Financial Times*, May 13, p. 21.
Haszeldine, S. 2012. UK carbon capture and storage, where is it? *Energy & Environment*, 23(2–3), 437–450.
Hickman, L. 2012. Is the UK right to invest in carbon capture technology? *Guardian.com*, April 3. https://www.theguardian.com/environment/blog/2012/apr/03/carbon-capture-storage-uk-government
HMG. 2003. *Our Energy Future – Creating a Low Carbon Economy*. London: HMSO.
Houlder, V. 2004. The case for carbon capture and storage. *Financial Times*, January 23, p17.
Hughes, C. 2012. CCS in the UK fears may be liars or may be not. *Energy & Environment*, 23(2–3), 413–416. https://doi.org/10.1260/0958-305X.23.2-3.413
Inderberg, T. and Wettestad, J. 2015. Carbon capture and storage in the UK and Germany: easier task, stronger commitment? *Environmental Politics*, 24(6), 1014–1033. https://doi.org/10.1080/09644016.2015.1062592
IPCC, 2005. Carbon Capture and Storage. Cambridge: Cambridge University Press.
Kemp, J. 2012. Desperation in the CCS gamble. *Climate Spectator*, April 4.
Littlecott, C. 2012. Stakeholder interests and the evolution of UK CCS policy. *Energy & Environment*, 23(2–3), 425–436. https://doi.org/10.1260/0958-305X.23.2-3.425
Macalister, T. 2007. BP scraps £500m Scottish carbon capture scheme. *The Guardian*, 25 May. www.theguardian.com/business/2007/may/25/oilandpetrol.news
Mason, R. 2010. Blow to British coal as Powerfuel hits rocks. *The Daily Telegraph*, December 10, p. 4.
McKie, R. 2005. Seabed supplies a cure for global warming crisis. *The Observer*, April 24, p. 12.

Mckie, R. 2007. Britain's lost opportunity to protect the planet. *The Observer*, June 17, p. 29.

Mckiernan, J. 2013. Red tape could lead to failure for proposed power project. *Aberdeen Evening Express*, November 4, p. 12.

McLachlan, C. 2003. Letter about CCS research. *The Guardian*, September 9. https://www.theguardian.com/environment/2003/sep/09/energy.renewableenergy

McLaughlin, M. and Bews, L. 2013. Ewing 'convinced' by CCS technology. *PA Newswire: Scotland*, October 10.

Monbiot, G. 2006. Heat: How we can stop the planet burning. Penguin UK.

Najor, P. 2005. 11 U.K. Companies unite to promote carbon capture, storage technology. *Greenwire*, 10(9).

NAO. 2012. *Carbon Capture and Storage: Lessons from the Competition for the First UK Demonstration*. London: National Audit Office.

Pearce, F. 2005. Squeaky clean fossil fuels; The big money's behind an ambitious plan to build zero-emission power stations that burn coal or gas. *New Scientist*, April 30, p. 26.

Pearce, F. 2008. Let's bury coal's carbon problem. *New Scientist*, 197(2649), 36–39. https://doi.org/10.1016/s0262-4079(08)60793-9

Perry, D. 2008. Ministers accused of dragging heels on energy technology. *Aberdeen Press and Journal*, July 22, p. 8.

Price, K. 2014. Teesside's carbon capture ambitions move forward. *Evening Gazette*, March 7, p. 12.

Prosser, D. 2007. Britain may have missed its chance for clean coal. *The Independent*, May 25.

Rowell, A. and Stockman, L. 2021. *Carbon Capture: Five Decades of False Hope, Hype, and Hot Air.* Oil Change International. https://priceofoil.org/2021/06/17/carbon-capture-five-decades-of-industry-false-hope-hype-and-hot-air/

Rustici, C. 2023. Carbon as waste and resource: The challenges of CCS and CCUS. *Direct Industry Magazine*, June 8. https://emag.directindustry.com/2023/06/08/carbon-capture-waste-and-resource-the-challenges-of-ccs-and-ccus-storage-transport-utilization/

Science and Technology Committee. 2006. *Meeting UK Energy and Climate Needs: The Role of Carbon Capture and Storage*. https://publications.parliament.uk/pa/cm200506/cmselect/cmsctech/578/578i.pdf

Science and Technology Select Committee. 2006. *CCS Enquiry*. London: HMSO.

Scrase, I. and Watson, J. 2009. CCS in the UK: Squaring coal use with climate change? In *Caching the Carbon: The Politics and Policy of Carbon Capture and Storage*, edited by Meadowcraft, J. and Langhelle, O. Cheltenham: Edward Elgar, pp. 158–185.

Sculthorpe, T. 2013. CCS fund 'will not be axed.' *Press Association Mediapoint*, June 6.

Shackley, S. and Evar, B. 2012. Up and down with CCS: the issue-attention cycle and the political dynamics of decarbonisation. In *The Social Dynamics of Carbon Capture and Storage* (pp. 149–171). London: Routledge.

Shell, 2002. Meeting the Energy Challenge. https://www.shell.com/sustainability/transparency-and-sustainability-reporting/sustainability-reports/_jcr_content/root/main/section/list/list_item/text.multi.stream/1657185127066/

bb64d059a8e3600ba9bad2b93db97008c7ea703d/shell-sustainability-report-2002.pdf
Smedley, M. 2005. BP to unveil major project. *World Gas Intelligence*, June 22.
Smith, L. 2010. Carbon capture and storage. In *Key Issues for the New Parliament*. House of Commons Library Research, pp. 100–101. House of Commons Research Library.
Stone, W 2010. Britain – E.ON out of carbon deal. *Morning Star*, October 21.
TUC/CCSA. 2013. *A UK Vision for Carbon Capture and Storage*. London: Trades Union Council.
US Department of State. 2003. *United States and European Union Joint Meeting on Climate Change Science and Technology Research*, February 10. https://2001-2009.state.gov/r/pa/prs/ps/2003/17493.htm
Vidal, J. 2008. Not guilty: the Greenpeace activists who used climate change as a legal defence. *The Guardian*, September 11. www.theguardian.com/environment/2008/sep/11/activists.kingsnorthclimatecamp
Warren, L. 2012. Where now for CCS in the UK? *Greenhouse Gas Science and Technology*, 2(3–5). https://doi.org/10.1002/ghg
Webb, T. 2007. Centrica will challenge decision on coal plant. *The Guardian*, October 14. https://www.theguardian.com/business/2007/oct/14/money.nuclearindustry
Wilson, K. 2007. Fury as pull-out dashes hopes of a thousand jobs. *Aberdeen Evening Express*, May 24, p. 2.
Wood, J. 2008. Environment: Court challenge may put back CCS project. *Utility Week*, November 14.
Zhongyang, L., Jiutian, Z. and Burnard, K. 2009. Near zero emissions coal: A China – UK initiative to develop CCS in China. *Energy Procedia*, 1(1), 3909–3916.

5 Third Time Lucky, 2015–2023

Contents

5.1 Introduction

The last chapter ended with a cliffhanger. In this chapter, all is explained. In the first section, the reasons for the cancellation of the second competition are given, and responses – both short-term and long-term – to it are explored. It is a story of remarkable resilience, coordination and determination, culminating in an unlikely renaissance for CCS in the UK exactly three years and three days after the cancellation. I tell it in four periods, named "shocked and overawed", "the price can be right," "the waiting game" and "back in the game". The obvious question – how was it done – is then explored (this material is covered in greater depth in Hudson and Lockwood (2023), and the account below draws on it and another related work (Hudson, 2023). Then a quick overview of the last four years of developments of CCS in the UK, happening in the context of the growth of "industrial clusters", is outlined. However, CCS has not moved merely from coal to gas to industrial functions. It is also being examined for its potential for "greenhouse gas removals" (akin to creating a way of creating a syphon from the bathtub, bypassing the plug and straight into the drain below). The history and policymaking around two

DOI: 10.4324/9781003461067-5

particular concepts – Bio-Energy Carbon Capture and Storage (BECCS) and DAC – are explored.

5.2 The "Kipling Manoeuvre": 2015–2018

Rudyard Kipling's *If* consistently tops polls and is the nation's favourite poem. It contains various pieces of advice for being phlegmatic – for meeting triumph and disaster and being able to "treat those two imposters just the same". For the purposes of this chapter, the crucial lines are these:

> if you can see the things you've given your life to broken
> And stoop and pick them up with worn out tools.[1]

This task – stooping and picking up – faced CCS advocates, some of whom had spent a decade or more battling for CCS and had already been through the disappointment surrounding DF1 and the prolonged first competition (covered in Chapter 4).

Before the story of the successful campaign to reassert CCS can be told, a note of caution about why the decision to cancel was made. It is important to guard against the assumption that consequential decisions or actions must have profound and well-considered causes. While this may be true, often the truth is more mundane – and it seems to be the case here. There seems to have been no great antipathy to CCS in David Cameron's government. Rather, the cause of the decision seems to have been more mundane – in the midst of seeking savings in the context of wanting to be able to free up money for other – more electorally salient spending, on police budgets and so on.

In 2017, a civil servant, Alex Chisholm, giving testimony to an inquiry by the House of Commons Public Accounts Committee, noted:

> It did reflect political decisions. . . . The Prime Minister at the time did actually say that: "Carbon capture and storage is £1 billion of capital expenditure – £1 billion that we could spend on flood defences, schools, or the health service".
>
> (Quoted in the House of Commons Committee of Public Accounts, 2017: 42; see also Tallentire (2015))

5.2.1 *November 2015 to March 2016: shocked and overawed*

In cancelling the second competition, the government chose not to call a press conference or even make a dedicated press release. Instead, the information was appended, almost as if it were an irrelevant afterthought, to another announcement to the Stock Exchange, on the afternoon of 25 November.

Four months later, speaking in a parliamentary debate, a Scottish National Party (SNP) MP, Callum McCaig, put it in these terms:

> I remember sitting on this very Bench, looking through the Budget statement and being somewhat relieved that the rumours I had heard about this competition being scrapped did not appear to be in that statement. Lo and behold, however, an announcement was made to the stock market a few moments after the Chancellor had left the Chamber, removing that funding.
>
> (Hansard, 16 March 2016)

Interviewed years later, someone who had been a senior civil servant at the time recalled:

> [I]t was a bad decision, in the sense that I'm not sure that . . . the competition design was perfect in terms of how would you get CCS deployed in the UK, but like, it wasn't just people having been walked up the garden path – they were at the door, you know, with their finger on the doorbell, and then their finger got chopped off. . . . I think a lot of ministers, frankly, felt that the anger was justified.
>
> (Hudson & Lockwood, 2023)[2]

No stranger to disappointment, given that it had been in existence since 2005, the CCSA was surprisingly (but also understandably) blunt in its assessment. Eschewing the usual conciliatory tone of previous post-disappointment communications, it released a press release on the afternoon of November 25, under the title ' "Chancellor Deals Devastating Blow to CCS Industry".

It included a quote by CEO Luke Warren: "Moving the goalposts just at the time when a four-year competition is about to conclude is an appalling way to do business" (CCSA, 2015).

This period, a few months long, was a mix of bewilderment and anger (Cozier, 2016). CCS advocates pulled on whatever levers were available to them. Ministers – and the prime minister himself – were quizzed at various meetings and events. In December, the Energy Minister Angela Leadsom was asked by the CCS Development Forum (an advisory body that had been established alongside the second competition in early 2012) to explain the government's position. Leadsom was also invited to a meeting of the All-Party Parliamentary Group on CCS. Leadsom's response was a classic of the non-committal response: "The Government continues to view CCS as having a potential role in the long-term decarbonisation of the UK's power and industrial sectors" (Macalister, 2016).

In late January 2016, a public event was held that captured the flavour of the time. It was a meeting organised by the UKCCSRC (a grouping of academics established in 2012) and held at Imperial College London under the title "Is CCS Dead, and if not, how do we resuscitate it?"

At the same time, an open letter was sent to Prime Minister David Cameron. It was signed by 12 organisations, including the CCSA, E3G, Energy Intensive Users Group, the Trade Union Congress and the UK Hydrogen Fuel Cell Association. It expressed "dismay and deep concern" about the cancellation and argued:

> Without an immediate, coherent and substantive response from Government to its recent policy reversals, confidence amongst project developers and investors across the full range of low-carbon technologies risks being irreversibly damaged. This would put at risk the fulfilment of legally-binding climate change targets, jeopardise the prospects for job creation and retention across carbon intensive sectors, and undermine the case for new investment in electricity generation capacity.

Labour MPs asked questions in the House of Commons, and one, Lisa Nandy, managed to initiate a National Audit Office investigation into the second competition and its cancellation.

In early January, the Energy and Climate Change Committee, which had several CCS- sympathetic MPs, announced an inquiry. By mid-February, it had held an evidence session and received written submissions and produced a report in near-record time. It found that the government needed urgently to recommit to CCS to restore investor confidence.

Simultaneously, the Teesside Collective sought meetings with ministers and used them to explain that the cancellation of the second competition had consequences for investor confidence.

SNP MPs tried to force the government to consult on a new CCS strategy, based on extensive stakeholder engagement, to be ready by June 2017 (Anon, 2016). They did this by tabling amendments to the Energy Bill before parliament. In response to this a government spokesperson said:

> We haven't closed the door to CCS technology in the UK, but as part of our ongoing work to get Britain's finances back on track, we have had to take difficult decisions to control government spending. CCS should come down in cost and we are considering the role that it could play in the long-term decarbonisation of the UK.
>
> (Harvey, 2016)

The period came to an end in February/March. By then, the advocates of CCS had come to realise that the only way forward for CCS was to combat the "price" argument that had been used against it.

5.2.2 April to September 2016: the price can be right

CCS advocates had to maintain both a sense of momentum (or at the very least possibility!) and also combat the price argument.

Three national level actors stepped forward – and one regional player appears to have been very important. The three nationals were unsurprisingly

- the CCSA, which faced an existential threat (it initially shed members, but its membership list recovered quickly),
- the CCC, which commissioned a report from a consultancy called Pöyry and created an expert group around it and
- the UKCCSRC, which held an emergency meeting in February 2016 and throughout the rest of that year a series of regional meetings.

The regional actor was the Teesside Collective, which had ironically come into existence because Teesside's bid in the second competition had been unsuccessful. It was well placed to make arguments about regional renaissance and industrial CCS, both of which would become more important after the 23 June 2016 "Brexit" referendum.

Even before the competition cancellation, there had been awareness of a need for further work on CCS's future. Shortly before the abrupt cancellation of the second competition, Lord Bourne, then a DECC minister, had asked Lord Oxburgh (then Chair of the CCSA) to undertake a study about the CCS's expansion. Now, instead of answering the question "how do we extend CCS?" the question became "how do we save it?" As Oxburgh admitted in his letter introducing the report:

> [A]fter so many false starts I began this study, as I know a number of my colleagues did, quite prepared to advise you to write-off CCS as a part of UK energy policy.

A number of other organisations were also working on reports in the spring of 2016. The CCC commissioned the consultancy firm Pöyry to investigate the question of how CCS could be deployed more cost-effectively as part of its work for the 2016 progress report (Pöyry, 2016). The CCSA undertook its own "lessons learned" review of the failed competitions (Dixon & Mitchell, 2016). Meanwhile, the TUC, in a report on the UK's green industries, also called for a relaunch of CCS (TUC, 2016).

The Oxburgh Report was published in September 2016. Its message was that CCS could be low-cost but that a new approach to policy support was needed.

5.2.3   October 2016 to mid-2017: the waiting game

The Oxburgh Report did not function like a slap, with the government nursing its cheek and saying, "Thanks, I needed that". Indeed, the launch was inauspicious. As one participant recalls, it was launched in a "*small, sweaty room somewhere in Westminster . . . and the audience was about 20 or 30 people or so. So the whole CCS endeavour had collapsed to that*" (cited in Hudson & Lockwood, 2023: 9).

In fact, there was a long period where it was not clear if the efforts had had any impact at all. There were warm words from ministers, emollient phraseology from civil servants and the usual platitudes. It was not clear what impact, if any, the arguments of the various groups had had around the essential and affordable nature of CCS.

At the time, a great deal of uncertainty had entered the picture because the UK had narrowly voted to leave the EU. The new prime minister, Theresa May, had made a focus on industrial renaissance part of her manifesto to become leader, and work was soon underway on a new industrial strategy. She combined two departments – DECC and BIS – which had previously looked at CCS for industrial decarbonisation – together as "Business, Energy and Industrial Strategy" (BEIS).

The proponents of CCS produced a constant flow of reports, conferences and lobbying. The main themes were twofold. First was the argument for a "hydrogen economy" (CCS would be needed to capture emissions from creating hydrogen from methane). Second was the goal of saving industries and jobs in the "industrial heartlands" – especially on the East Coast of England. Their cause was helped by the arrival of a Conservative, Ben Houchen, as Mayor of the traditionally Labour area of Teesside in 2017. Houchen quickly began to amplify the arguments of groups like the Teesside Collective and the UKCCSRC that CCS was both affordable and necessary.

The UKCCSRC had held a series of regional meetings and workshops about the importance of CCS to regional economies (UKCCSRC, 2016a, 2016b, 2016c). These arguments are captured well in this quote:

[C]arbon capture and storage (CCS) is currently the only option for achieving deep reductions from some industries, such as petrochemicals, cement, and refining. Major industrial clusters at Teesside and Grangemouth each produce large volumes of CO_2. With CO_2 capture technology in place, each cluster could handle several million tonnes of CO_2 a year avoiding emissions and helping industry achieve decarbonisation targets. Availability of CCS infrastructure enables large-scale, low-carbon hydrogen supply allowing decarbonisation of 'hard to treat' transport and heat sectors.

(UKCCSRC, 2016a: 1)

The "local angle" also helped to create political support for CCS. As one advocate (quoted in Hudson & Lockwood, 2023) said:

[Y]ou need to move CCS from a 'technology with no mates' to a technology which local councillors regard is important. And . . . MPs get asked questions. As MPs are passing through the lobby, we need people to ask questions to their MP, "what about our CCS project?" And so it becomes much more locally-owned and supported amongst a diversity of people, rather than the few relatively remote technology advocates.

In January 2017, an Industrial Strategy Green Paper was published. While this did not contain any mention of CCS, clean energy was a major theme. The government committed to commissioning "a review of the opportunities to reduce the cost of achieving our decarbonisation goals in the power and industrial sectors" (HMG, 2017: 94). Meanwhile, the Autumn Statement of November 2016 had signalled funding for an Industrial Strategy Challenge Fund to provide R&D funding. This Fund was announced in April 2017 and launched in May. The stars were beginning to align.

In June, a new minister entered the fray. Claire Perry replaced Nick Hurd at BEIS. Hurd had been well-respected. As one interviewee recalled:

> [W]e were really, really sad to see Nick Hurd leave and Claire Perry come in, because we thought, "well, there's a risk here, she may not be kind of particularly interested in CCS". But in fact, Claire Perry was just absolutely brilliant and kind of picked this up and wanted to be very ambitious in this space. She drove it forward.

Perry, interviewed by the author, said:

> [I]t felt like we did a lot in quite a short period of time. And we had on the stocks, the clean growth plan. This kind of, you know, we were going to launch a plan, which sort of, you know, had been written and had been written by committee and you know, these things. And essentially, I rewrote it that summer, personally, with the help of Guy Newey and some fabulous civil servants, because it seemed, and I have a, you know, a background in business and strategy. And it seemed to me that you were, we were talking about a strategy not a plan. Because what we needed to set out was that where do we actually want to get to? What are the pieces of this puzzle? Some things were further advanced than others; we could actually say, "we're going to do XYZ".
>
> And my view, once I got into it was, as we started to talk about Net Zero, we've got to remove carbon. We just have to, and I understand the ideologically, this, there was really strong opposition to that concept. You know, people wanted to explore mitigation or reduction to the nth degree, but ultimately, you had to remove and you can see that the IPCC kind of came around to that point of view. And what was the best way to do it. And also, the conversation was not just about decarbonisation of the energy system, it was the hard-to-abate stuff. So yes, we could wait for hydrogen to drop in price and for [new solutions] to come along for the steel sector. But we sure as hell needed to extract from carbon more quickly than that. And that was the genesis of the cluster idea.

Perry's immediate focus was the Clean Growth Strategy (CGS). This was the statutory response to the government's fifth Carbon Budget. It was already long overdue when she came into office. While the CGS did not make overt commitments, the language was extremely encouraging.

> We now see a new opportunity for the UK to become the global technology leader for CCUS, working internationally with industry and governments to bring about global cost reductions. We will do this through: Re-affirming our commitment to deploying CCUS in the UK subject to cost reduction:
>
> We will . . . convene a CCUS Cost Challenge Taskforce to deliver a plan to reduce the cost of deploying CCUS. This will then underpin a deployment pathway for CCUS in 2018, setting out the steps needed to meet our ambition of deploying CCUS at scale during the 2030s, subject to costs coming down sufficiently.
>
> Following the advice from the Parliamentary Advisory Group on CCUS (the Oxburgh Report) the Government will review the delivery and investment models for CCUS in the UK to understand how the barriers to deployment can be reduced, and how the private and public sectors can work together to deliver the Government's ambition for CCUS.
>
> (HMG, 2017: 70)

5.2.4   October 2017 to November 2018: back in the game

The final period here – from October 2017 to November 2018 – is really a consolidation – and extension – of the gains made during 2017. The key forum here is the *Cost Challenge Taskforce*. Several people interviewed by the author regard it as pivotal. One saw it as marking:

> The start of an exponential increase . . . was sort of establishment reset, – the official reset. Here, we had a workable plan, which BEIS could get behind and feel was deliverable in the language or the terminology terms of reference, which British government would understand. Because it included industry buy-in, it included cost reduction, and redesign of the delivery model for CCS away from the A to B onto this "A to B plus C", where there would be a transport and storage operator.

In addition, the Taskforce also more deeply embedded industrial decarbonisation within the rationale for CCS. Another participant said that it "[e]nlarged [and] reframed the issue as one of industrial clusters and industrial decarbonisation. So, it's been kind of whatever the word is, sawtooth profile, there's something quite extreme". Central to the workings of the Taskforce was a new, leading role for industry in shaping policy. Two quotes illustrate this:

> [W]e saw this as a tremendous shift a few years ago, from if you like, public authorities, talking about it and hoping to do something and the Government

commissioning this, that and the other to the industry, to industry taking the lead. So your British Gas, and that's National Grid and others; they started to take the lead to say "the only way forward is if we do this". I would think that the flip took place, I would say, about three to four years ago.

and

[T]here was the Lord Oxburgh group, and then the Cost Challenge Task Force, that thing. But at the same time, you know, the industry itself, you know, I think had been challenged by the failure of this competition to look at, well, what went wrong? And why, and what can we do about that? So I think, actually, I think it was a really good sort of close collaboration between government and industry to really test some of these ideas and understand how you might deliver them.

In his testimony to the BEIS Committee in November 2018, Ashley Ibbett, BEIS Director of Clean Growth, noted that BEIS "*had tremendous interest and enthusiasm from industry to a level we did not really anticipate*" (BEIS Committee, 2019: 34).

There is of course another way of seeing this – that of capture of state decision-making processes by vested interests. Pointing to the secretariat functions performed for the Cost Challenge Taskforce by Linklaters, a law firm with close links to BP, Ahmed (2021) argued:

The relationship between BP, Linklaters and the UK Government raises serious conflict of interest questions. These relate to BP's role as a major client of Linklaters, BP's input into the work of the taskforce chaired by Linklaters, and BP's resulting windfall thanks to two Government-backed multibillion-pound CCUS schemes.

The Cost Challenge Taskforce produced its report in mid-2018. All that remained was for an official endorsement of the approach and accompanying funding. This came months later. On 28 November, the government released its deployment action plan (HMG, 2018). This action plan adopted most of the recommendations of the Taskforce.

On the same day, almost exactly three years since the traumatic and potentially fatal cancellation of the second competition, Minister Claire Perry stood at a lectern addressing an international conference on CCS in Edinburgh. She put the view that

the UK is setting a world-leading ambition for developing and deploying carbon capture and storage technology to cut emissions. It shows how determined all countries are to unlock the potential of this game-changing technology that representatives from across the globe are gathered here today in Edinburgh. The time is now to seize this challenge to tackle climate change while kick starting an entirely new industry.

Funding was also pledged, with £20 million for supporting CCUS technologies at industrial sites across the UK and £315 million for decarbonising industry, including the potential to use CCUS.

In December, at the COP meeting in Poland, Perry also announced another £170 million in funding for the development of a zero-carbon industrial hub. Further funding pledges were to follow, including £1 billion for a CCUS Infrastructure Fund in November 2020.

From the seeming catastrophe and the drying up of the streams in November 2015, advocates of CCS had managed to re-open a policy window in three years and three days.

To frame this in the language of the MSA, we can see that a number of factors are combined in the three streams, with policy entrepreneurs acting forcefully to couple those streams and open a window (Figure 5.1).

5.2.4.1   The window opens

There was no single factor or heroic individual responsible for CCS's return. Nor was its comeback inevitable. Those who thought CCS essential (either for their industry or their region or for global emissions reductions) did a variety of things. They engaged as follows:

- Worked assiduously to argue that the price would not be as high as the gloomiest predictions
- Built a much bigger network of advocates (especially regionally)

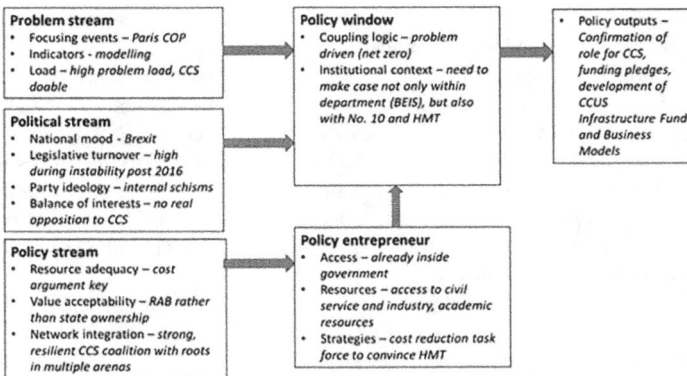

Figure 5.1 The policy window reopened

- Used the venues that were available to help "co-curate" policy development
- Tied the technology to a narrative of industrial growth and innovation, and regional economic development and jobs
- Tied the technology to the renewed enthusiasm for hydrogen.

Hudson (2023)

5.2.5   *2019–2023: progress, but slower than anticipated*

Having dealt with the comeback, the coverage of the last four years can be undertaken. It is important to remember that these have been a period of extraordinary turbulence for the UK, with its exit from the EU, a pandemic, a carousel of prime ministers, energy ministers, departmental reorganisations and, latterly, a war in Europe.

While there is remarkable regional enthusiasm for the idea of "industrial clusters", hinging on carbon capture and storage projects, the fine detail of the business models that would support this, and allow investors to make FIDs, is still not in place. Meanwhile, the "leading position" for CCS seems to have shifted, in the aftermath of President Biden's climate change stimulus package, especially the Inflation Reduction Act, to the United States.

Although the November 2018 CCUS Action Plan made a splash and acted as a rallying point, there remained doubt about its likely efficacy. In words that seem more prescient than ever, with the passage of a further four years, a select committee of MPs wrote in April 2019:

> Whilst we strongly support cost minimisation, we disagree with the CCUS Action Plan's stipulation that deployment "at scale" should be supported only if "sufficient" cost reductions are achieved. Such vague terminology gives no certainty to investors and does little to ensure that CCUS can contribute to meeting the UK's overarching climate change targets at least cost, given its existing status as the cheapest – or only – decarbonisation option in many industrial applications.
>
> We recommend that Government revise its formal aims in light of the Minister's more nuanced position and prioritise the development of clear ambitions that will bolster its renewed efforts to kickstart CCUS.
>
> We further recommend that the Government commits to supporting CCUS where and whilst it remains the cheapest route to decarbonisation, notably in industrial applications. Rather than seeking unspecified cost reductions, the Government should set out plans to ensure that projects are brought forwards at least cost.
>
> (BEIS Committee, 2019: 13)

While the MPs were saying this, a CCS Advisory Group was busy producing detailed advice about various funding models. This has now split into industrial and power-generation models, and there are ongoing (for several years) rounds of consultations, first under the auspices of BEIS and more recently under the rebranded Department of Energy Security and Net Zero.

In March 2021, the Industrial Strategy that had been such an important underpinning for the pivotal CGS was abandoned and replaced with a Treasury document. It was "a move that even surprised some of those working in Boris Johnson's government at the time". "The Treasury blindsided us", said one official. The strategy was ditched in favour of a "Build Back Better" growth plan. Kwasi Kwarteng, the free-market former Tory business secretary at the time, described Clark's strategy as "a pudding without a theme" (Parker et al., 2023).

In the same month, an "Industrial Decarbonisation Strategy" was released. In October 2021, two clusters were chosen – the East Coast Cluster, combining Humber and Teesside, and one in the North West, Hynet (Thomas, 2021).

This was one of many announcements in the lead up to the UK hosting COP26 in Glasgow. One outcome was

> the Glasgow Climate Pact; although 23 countries newly committed to phasing out coal power (bringing the total to 45), the final agreement was only to "phase down [unabated] coal power", making CCS important for countries still dependant on coal to meet climate targets. CCS is included in the 2030 Nationally Determined Contributions (due to be updated in 2022) of 11 countries, not including the UK (GCCSI, 2020). The Glasgow Climate Pact also included new Mission Innovation platforms to accelerate deployment of Carbon Dioxide Removal technologies and Net Zero Industries.
>
> (Gough & Mander, 2022)

Shortly after this, the government gave planning approval to a new coal mine in Cumbria. The defenders of this decision point to the fact that the coal extracted is for steel-making rather than power generation. They also used the existence of carbon capture and storage as a way of allaying concerns (Hudson, 2022). In November 2023, Adair Turner, who had been the first chair of the Climate Change Committee, stated it had been

> a disaster for our reputation, and it provides arguments for the people within government or within interest groups in China and India to say "oh look, the UK is supposedly committed to net zero, but it's not serious, it's building a new coal mine". And the same occurs with new oil and gas fields in the North Sea.
>
> (Harvey, 2023)

The same technique – of coupling a controversial decision with the invocation of CCS as a salve to otherwise-guilty consciences – was on display in

July 2023 when the government made two announcements on the same day, with the clear intention of muddying the waters. It announced two more CCS projects – the Viking Project on Humberside and (at long last, almost 20 years after Miller Field) a Scottish project, Acorn in Aberdeenshire – as a way of deflecting criticism of its second announcement – issuing of more oil and gas exploration licences for the North Sea.

While these decisions have been the focus of much protest and attention, the legislative process grinds on. On 7 April 2022, the "British Energy Security Strategy" (HMG, 2022a) was released, with the tagline "secure, clean and affordable British energy for the long term". Three months later, in early July, an "Energy Security Bill" (since renamed the "Energy Bill") was introduced into the House of Lords. The bill, which its authors called the largest piece of primary legislation concerning energy since the Energy Act 2013, had the aim of providing the necessary legislative foundations for key policy measures around energy security and transition. Its focus was three areas:

- Leveraging of clean technologies investment
- Reform of the UK's energy system and protection of consumers
- Maintenance of the safety, security and resilience of the energy systems across the UK

This occurred under Prime Minister Boris Johnson. After his resignation, and that of his short-lived successor Liz Truss, the Bill was reviewed and updated in December 2022. On 25 April 2023, the Energy Bill was introduced in the House of Commons.

As Sheppard et al. (2023) reported, Ruth Herbert, chief executive at the CCSA, welcomed the announcement:

"It's really great to have this momentum but there is still a huge amount to build by 2030", she said, referring to the government's target of capturing between 20mn and 30mn tonnes of CO_2 per year by the end of the decade. "Billions of pounds of investment is waiting to be deployed to decarbonise these industrial regions, but firm plans are required to secure it".

The Energy Bill gained Royal Assent (and became the Energy Act) in late October 2023.

Its passage was warmly welcomed by pro-CCS organisations. Olivia Powis, UK Director at the *CCSA*, said:

The CCSA welcomes Royal Assent for the Energy Bill, which provides the enabling legislation for CCUS business models, and builds on significant progress for the technology this year.

Those with a knowledge of just how long and hard the CCSA had been pushing for this could be forgiven for inferring a small amount of teeth-gritting and weariness in her following sentences.

To fully capitalise on this opportunity, we encourage the government to commit to timely cluster delivery, a transparent deployment plan to 2035, streamlined permitting processes, a robust supply chain and enhanced public support. Measures in the new Energy Act will enable us to unlock the full potential of CCUS and further advance our nation toward a cleaner, more sustainable future, helping to ensure the UK's Green Economy has the opportunity to lead the next Industrial Revolution.

Aniruddha Sharma, CEO and Chairman of *Carbon Clean*, welcomed the passage of the Energy Act, a milestone for meeting the UK's decarbonisation goals, creating a legal framework for the support schemes needed to make carbon capture a reality.

The Act also contains key provisions to regulate the future carbon capture, utilisation and storage (CCUS) industry, which will help ensure the sector delivers value for money as it grows and maintains public support. The Act lays solid foundations, and the challenge now, for industry and policymakers alike, is to build on these foundations and scale-up the UK's CCUS sector as quickly as possible.

(Both cited in George, 2023)

However, the Energy Act is most emphatically not the delivery of the (long-delayed, long-awaited) necessary detail for investment decisions.

It is important to note that while the EA 2023 has laid the foundations for a new regulatory framework, the finer details of that framework will be forthcoming through secondary legislation.

(Hedges & Thomson, 2023)

The EA 2023 is just the tip of a regulatory iceberg, and a large body of secondary legislation will need to be developed fast to make a real impact on the market and ensure that the UK does not fall behind in the global hydrogen and CCUS race.

Given that on the same day the Energy Act became law the Energy Secretary Clare Coutinho refused to update the government's Carbon Budget Delivery Plan in light of the significant backtracking by the Sunak administration over the summer (Blackman, 2023), it is fair to expect that there will be further delays and difficulties.

5.2.6   What is taking so long?

When conducting interviews in mid-2022, the author asked various close observers and participants for their view on the speed of the process in decision-making. While some (especially in industry and consultancy) expressed exasperation and frustration, others, with long experience of the internal workings of government, had a slightly different perspective. They pointed to the fact that the government had had a great deal of unprecedented pressures on it, around the pandemic, and pointed to the fact that, despite all this, CCS was still "on the agenda" and had not disappeared in the way that the notion of "levelling up" largely had. Of course, one of the key arguments that CCS advocates deploy is that it will protect – and even create – new jobs. The main locus of attention was not on civil servants, who were regarded as diligent and proactive, but rather on ministers and Secretaries of State who were rotating through relevant departments at high speed.

In its annual report to parliament for 2023, the CCC had noted:

> Due to high capital and operating costs, together with technology and construction risks associated with novel technologies, engineered removals will require Government support in their early stages of deployment. Clarity on the form that this support will take is now overdue. The Government's response to its Power BECCS Business Models consultation in March 2023 shows progress, but a response to the Engineered Removals Business Models consultation is needed this year to ensure other approaches are able to progress.
>
> (CCC, 2023: 323)

The bulk of this chapter has dealt with power and industrial CCS, with the latter taking a much more prominent role than it previously had. The chapter closes with consideration of a third strand to CCS – that of greenhouse gas removals.

5.5   Greenhouse gas removal

For a very long time after the 1988 "discovery" of the climate issue, the focus of policymakers was on decarbonising – slowly or quickly – energy systems and perhaps transport. This was unsurprising, given that for the last century they have largely relied on the (ever-expanding) use of fossil fuels. As time passed, it became obvious that merely addressing energy and transport (which, to be clear, has still not been done) would not be enough to reduce emissions at the speed required. More attention began to be paid not just to industry – especially cement and steel – but to other sectors such as food production and aviation.

The idea of removing greenhouse gases via artificial means was already being spoken over 20 years ago. BECCS was first publicly mooted in 2001 (Obersteiner et al., 2001). CCS was one of the so-called stabilisation wedges of Pacala and Socolow (2004). However, it is fair to say that it is only after the pivotal 2015 COP in Paris that serious attention began to be paid to the question of removing greenhouse gases from the atmosphere so that not merely emissions flowing into the bathtub might be reduced to an alleged trickle, but the level of water in the bathtub itself might be lowered, thanks to the removal of carbon dioxide already in the atmosphere. The task of capturing and then storing so much carbon dioxide as to actually reduce atmospheric concentrations is not on any prospectus at present.

It is important to remember that notes of caution about reliance on GGR were already being sounded at this time. Two academics conducted interviews with experts and found that "Unrealistically optimistic assumptions regarding the future availability of BECCS in IAM scenarios could lead to the overshoot of critical warming limits and have significant impacts on nearterm mitigation options".

(Vaughan & Gough, 2016)

However, that same year the CCC advised:

Even with full deployment of known low-carbon measures some UK emissions will remain, especially from aviation, agriculture and parts of industry. Greenhouse gas removal options (e.g. afforestation, carbon-storing materials, bioenergy with carbon capture and storage, and direct air capture and storage) will be required alongside widespread decarbonisation in order to reach net zero emissions.

(CCC, 2016: 7)

In October 2018, the UK's Royal Society and Royal Academy of Engineering issued a joint report on GGR (Royal Society, 2018; Harvey, 2018). It argued that the UK could have a carbon-neutral economy during the second half of this century if strong curbs on fossil fuels were accompanied by widespread deployment of CCS.

Months later, in its report on Net Zero, the Committee on Climate Change indicated that GGR technologies and their associated value chains would be required to generate negative emissions of up to 60–90 $MtCO_2$ per year by 2050 to meet national emissions targets cost-effectively (CCC, 2019).

The following year saw two crucial developments around Greenhouse Gas removals.

First, in the middle of the year, Direct Air Capture gained a huge boost.

On 30 June, in a speech called *A New Deal for Britain*, Prime Minister Boris Johnson announced up to £100 million of new research and development funding to help develop direct air capture technologies in the UK. According to a report in *the Times*:

> [T]he project had been driven by Boris Johnson's chief adviser [Dominic Cummings], who has become an advocate for the technology and its potential in the battle against climate change. He has won the support of Tim Leunig, the chancellor's economic adviser, who is also backing it.
>
> Both men believe that with early investment Britain could become a world leader in the technology, which is only being developed by two firms. . . .
>
> "Dom had become obsessed by this", a Whitehall source said. "He's the one who has been pushing it despite huge scepticism from officials. But he's got his way".
>
> (Wright, 2020)

The article also notes that some civil servants in Whitehall worried that a focus on DAC could "*distract from more conventional and proven projects to cut emissions such as the government's pledge to spend £9 billion on insulating homes*".

A competition has been launched, with 15 projects shortlisted in July 2022 (HMG, 2022). Academic attention is being paid to how DAC is perceived (Cox et al., 2020) and also how it might be governed (Sovacool et al., 2023).

Second, in October a newly formed entity, the *Coalition for Negative Emissions*, wrote to the then-Chancellor Rishi Sunak. The coalition, which included the Confederation of British Industry (CB), Velocys and Drax, was looking for the government to commit to GGR. It said:

> Together, we represent hundreds of thousands of workers across some of the UK's most critical industries, including aviation, energy and farming, each of which contribute billions of pounds each year to the economy.
>
> Whilst we should seek to decarbonise sectors such as aviation, heavy industry and agriculture as far as practically possible, due to technical or commercial barriers it is unlikely we will eliminate their greenhouse gas emissions completely. Negative emissions technologies are critical, therefore, to balancing out these residual emissions and ensuring we achieve Net Zero in a credible, cost effective and sustainable way.
>
> (Anon, 2020)

While correlation is not causation, it is noteworthy that the following month the government asked the National Infrastructure Commission (NIC) to provide recommendations on the following:

- The technologies that should be deployed to remove greenhouse gases from the atmosphere and deliver negative emissions

- The policies needed to incentivise their roll-out
- The timeline of decisions needed by government to enable the UK to use engineered removals to achieve net zero

The NIC delivered its report in July of the following year (NIC, 2021). It argued:

> Removing carbon dioxide from the air is expensive. This new sector could have revenues of £2 billion by 2030, and in the tens of billions by 2050. In the long term, polluting industries, not taxpayers, should bear these costs, paying for the engineered removals they need in a competitive market. But the costs should be phased in over time, and vulnerable or disadvantaged groups in society should be protected.

The NIC report also argued that in the near term, the government would "need to support the initial deployment of a portfolio of engineered removals, using policy mechanisms to bring providers to commercial readiness. In time, government should support the transition to a competitive market, which will be the most efficient solution", and made a series of specific recommendations.

5.6 Conclusion

In 2015, when this chapter began, the atmospheric concentration of carbon dioxide – the thickness of the blanket – was 401 ppm. The quantity of human emissions entering the atmosphere – water coming from the tap into the bathtub – was in the order of 50 billion tonnes. In 2023, we are at 42 ppm and 54 billion tonnes respectively, when scientists have been saying that the only way to avoid disaster is a radical reduction, globally, of human emissions.

CCS was first proposed as part of the solution to the build-up of greenhouse gases almost half a century ago. It was first mentioned as a possibility for the UK in 1989. After being largely ignored in the 1990s, as electricity privatisation took hold, it finally broke through in the period 2003. Now, after 20 years of proposed projects, a competition that fizzled out, a competition that was cancelled at the last minute, a recovery process and a new competition that expands the uses to which CCS might be put, the technology seems to be gaining attention and allies. Whether this is another false dawn, or the beginning of something new will be obvious in hindsight. In the final chapter I offer some speculations and advice for those interested in tracking what does, and does not, happen.

Notes

1 For an assessment of the poem, see Allen (2015).
2 This interview and others quoted below were part of an IDRIC-funded project that the author was funded by. The report can be found on the IDRIC website. An academic article has been submitted.

References

Ahmed, N. 2021. Carbon captured: The links between BP's law firm, a government taskforce, and billions in public contracts. *Byline Times*, November 18. https://bylinetimes.com/2021/11/18/carbon-captured-the-links-between-bps-law-firm-a-government-taskforce-and-billions-in-public-contracts/

Allen, A. 2015. Iffy: Behind the mask of Rudyard Kipling's confidence. *Poetry Foundation*. www.poetryfoundation.org/articles/70303/iffy

Anon. 2016. Shadow Energy Minister calls for national CCS strategy. *Edie*, February 23. www.edie.net/shadow-energy-minister-calls-for-national-ccs-strategy/

Anon. 2020. New coalition calls on Government to act on negative emissions technologies. *Biofuels International*, October 12. https://biofuels-news.com/news/new-coalition-calls-on-government-to-act-on-negative-emissions-technologies/

BEIS Committee. 2019. *Third Time Lucky*. BEIS, 2019. Carbon capture usage and storage: third time lucky? London: HMSO. https://publications.parliament.uk/pa/cm201719/cmselect/cmbeis/1094/1094.pdf

Blackman, D. 2023. Coutinho refuses to update carbon budget plan as Energy Act is passed. *Utility Week*, October 26. https://utilityweek.co.uk/coutinho-refuses-to-update-carbon-budget-plan-as-energy-bill-is-passed/

CCC. 2016. *UK Climate Action Following the Paris Agreement*. www.theccc.org.uk/wp-content/uploads/2016/10/UK-climate-action-following-the-Paris-Agreement-Committee-on-Climate-Change-October-2016.pdf

CCC. 2019. Net Zero: The UK's contribution to stopping global warming Climate Change Committee, https://www.theccc.org.uk/wp-content/uploads/2019/05/Net-Zero-The-UKs-contribution-to-stopping-global-warming.pdf

CCC. 2023. *Progress in Reducing UK Emissions: 2023 Report to Parliament*. www.theccc.org.uk/wp-content/uploads/2023/06/Progress-in-reducing-UK-emissions-2023-Report-to-Parliament.pdf

CCSA, 2015. Media Release: Chancellor Deals Devastating Blow to CCS Industry. London: CCSA.

Cox, E., Spence, E. and Pidgeon, N 2020. Public perceptions of carbon dioxide removal in the United States and the United Kingdom. *Nature Climate Change*, 10, 744–749. https://doi.org/10.1038/s41558-020-0823-z

Cozier, M. 2016. Reactions to the UK's cut to CCS funding: does CCS have a future in the UK? *Greenhouse Gases: Science and Technology*, 6(1), 3–6.

Dixon P, Mitchell T. 2016. Lessons Learned: Lessons and evidence derived from UK CCS Programmes, 2008–2015. Carbon Capture and Storage Association.

GCCSI, 2020. Global Status of CCS 2020. Global CCS Institute, Melbourne, Australia. https://www.globalccsinstitute.com/resources/global-status-report/download/

George, V. 2023. UK Government passes new energy Act 2023 To reduce bills. *Carbon Herald*, October 27. https://carbonherald.com/uk-government-passes-new-energy-act-2023-to-reduce-bills/

Gough, C. and Mander, S. 2022. CCS industrial clusters: Building a social license to operate. *International Journal of Greenhouse Gas Control*, 119.

Hansard. 2016. *House of Commons Debates*, March 14. https://publications. parliament.uk/pa/cm201516/cmhansrd/cm160314/debtext/160314-0002. htm

Harvey, F. 2016. Scrapping carbon capture support 'threatens UK climate targets'. *The Guardian*, February 10. www.theguardian.com/environment/2016/feb/10/scrapping-carbon-capture-support-threatens-uk-climate-targets

Harvey, F. 2018. Urgent greenhouse gas removal plan could see UK hit 'net zero' target – report. *The Guardian*, September 12. https://www.theguardian.com/environment/2018/sep/12/urgent-greenhouse-gas-removal-plan-could-see-uk-hit-net-zero-target-report

Harvey, F. 2023. Allowing Cumbria coalmine was 'disaster' for climate diplomacy, says Lord Turner. *The Guardian*, November 16. www.theguardian.com/environment/2023/nov/16/allowing-cumbria-coalmine-was-disaster-for-climate-diplomacy-says-lord-turner

Hedges, A. and Thomson, P. 2023. United Kingdom: Energy Act 2023 – impact on the hydrogen and carbon capture, utilisation and storage (CCUS) markets. *Global Compliance News*, November 13. www.globalcompliancenews.com/2023/11/13/https-insightplus-bakermckenzie-com-bm-energy-mining-infrastructure_1-united-kingdom-energy-act-2023-impact-on-the-hydrogen-and-carbon-capture-utilization-and-storage-ccus-market_11082023/

HMG. 2017. *The Clean Growth Strategy: Leading the Way to a Low Carbon Future*. https://assets.publishing.service.gov.uk/government/uploads/system/uploads/attachment_data/file/664563/industrial-strategy-white-paper-webready-version.pdf

HMG, 2018. *Clean Growth: The UK Carbon Capture Usage and Storage deployment pathway An Action Plan*. https://assets.publishing.service.gov.uk/media/5bfd760bed915d118adbb940/beis-ccus-action-plan.pdf

HMG 2022a. *British Energy Security Strategy*. https://assets.publishing.service.gov.uk/media/626112c0e90e07168e3fdba3/british-energy-security-strategy-web-accessible.pdf

HMG. 2022b. *British Energy Security Strategy*, April 7. www.gov.uk/government/publications/british-energy-security-strategy/british-energy-security-strategy

House of Commons Committee of Public Accounts. 2017. Carbon capture and storage. *Sixty-Fourth Report of 2016–17, HC1036*. https://publications.parliament.uk/pa/cm201617/cmselect/cmpubacc/1036/1036.pdf

Hudson, M. 2022. Cumbria coal mine: empty promises of carbon capture tech have excused digging up more fossil fuel for decades. *The Conversation*, December 8. https://theconversation.com/cumbria-coal-mine-empty-promises-of-carbon-capture-tech-have-excused-digging-up-more-fossil-fuel-for-decades-196242

Hudson, M. 2023. How carbon capture and storage was brought back from the dead, and what happens next. *Sussex Energy Group Blog*. https://blogs.sussex.ac.uk/sussexenergygroup/2023/01/30/how-carbon-capture-and-storage-was-brought-back-from-the-dead-and-what-happens-next/

Hudson, M. and Lockwood, M. 2023. *REPORT: Dead and unburied: The resurrection of carbon capture and storage in the UK 2015–2018*. *IDRIC*.

https://idric.org/resources/report-dead-and-unburied-the-resurrection-of-carbon-capture-and-storage-in-the-uk-2015–2018/

Macalister, T. 2016. Spending watchdog to examine scrapping of £1bn carbon capture plan. *The Guardian*, January 31. www.theguardian.com/business/2016/jan/31/spending-watchdog-nao-george-osborne-carbon-capture-storage-scheme

NIC. 2021. *Engineered Greenhouse Gas Removals*. https://nic.org.uk/app/uploads/NIC-July-2021-Engineered-Greenhouse-Gas-Removals-UPDATED.pdf

Obersteiner, M., Azar, C., Kossmeier, S., Mechler, R., Moellersten, K., Nilsson, S. . . . and Yan, J. 2001. Managing climate risk. *Science*, 294(5543), 786–787. https://doi.org/10.1126/science.294.5543.786

Pacala, S. and Socolow, R. 2004. Stabilization wedges: solving the climate problem for the next 50 years with current technologies. *Science*, 305(5686), 968–972.

Parker, G., Pfeiffer, F. L. and Shrimsley, R. 2023. Political Fix podcast episode: 'does Rishi Sunak have an industrial strategy?' *Financial Times*. www.ft.com/content/0fb0fee2-f16b-4691-b61f-ceda00e317e2

Pöyry, 2016. A Strategic Approach for Developing CCS in the UK. *Climate Change Committee* https://www.theccc.org.uk/wp-content/uploads/2016/07/Poyry_-_A_Strategic_Approach_For_Developing_CCS_in_the_UK.pdf

Royal Society and Royal Academy of Engineering. 2018. *Greenhouse Gas Removal*. https://royalsociety.org/-/media/policy/projects/greenhouse-gas-removal/royal-society-greenhouse-gas-removal-report-2018.pdf

Sheppard, D., Millard, R. and Parker, G. 2023. Industry calls on UK to accelerate carbon capture as new projects approved. *Financial Times*, July 31. https://www.ft.com/content/d2e46191-282b-4718-af1f-1cfedfbf7d23

Sovacool, B., Baum, C., Low, S., Roberts, C. and Steinhauser, J. 2023. Climate policy for a net-zero future: ten recommendations for Direct Air Capture. *Environmental Research Letters*, 17(7). https://doi.org/10.1088/1748–9326/ac77a4

Tallentire, M. 2015. More police cash and tax credits U-turn welcomed, but has Chancellor 'all but ignored' Teesside? *The Northern Echo*, November 25. www.thenorthernecho.co.uk/news/14103141.police-cash-tax-credits-u-turn-welcomed-chancellor-all-ignored-teesside/

Thomas, N. 2021. Carbon capture schemes speeded up; Global warming Two projects gain priority but Scots say selection has been 'completely illogical'. *Financial Times*, October 20, p. 2.

TUC, 2016. Powering Ahead: How UK industry can match Europe's environmental leaders. *Trades Union Council*, https://www.tuc.org.uk/research-analysis/reports/powering-ahead-how-uk-industry-can-match-europes-environmental-leaders

UKCCSRC, 2016a. Delivering Cost Effective CCS in the 2020s – a new start. https://ukccsrc.ac.uk/wp-content/uploads/2022/03/delivering_cost_effective_ccs_in_the_2020s_-_a_new_start_web_11.3.16.v2_final.pdf

UKCCSRC 2016b. Delivering Cost Effective CCS in the 2020s – an overview of possible North West developments. UKCCSRC, https://ukccsrc.ac.uk/wp-content/uploads/2022/03/delivering_cost_effective_ccs_in_the_2020s_an_overview_of_possible_north_west_developments_web.pdf

UKCCSRC 2016c. Delivering Cost Effective CCS in the 2020s: an overview of possible developments in Wales and areas linked to Welsh CCS activities via shipping. https://ukccsrc.ac.uk/wp-content/uploads/2022/03/delivering_cost_effective_ccs_in_the_2020s_an_overview_of_developments_linked_to_welsh_ccs_activities_via_shipping_web.pdf

Vaughan, N. E. and Gough, C. 2016. Expert assessment concludes negative emissions scenarios may not deliver. *Environmental Research Letters*, 11(9), 095003.

Wright, O. 2020. Dominic Cummings wins £100m to save planet by sucking CO_2 from air. *The Times*, July 3. https://www.thetimes.co.uk/article/dominic-cummings-wins-100m-to-save-planet-by-sucking-co2-from-air-8qv3mzjx8#:~:text=Dominic%20Cummings%20wins%20%C2%A3100m%20to%20save%20planet%20by%20sucking%20CO2%20from%20air,-Oliver%20Wright&text=An%20experimental%20plan%20championed%20by,100%20million%20from%20the%20Treasury.

6 Conclusion

Contents

6.1 Introduction

Recently, someone wrote the following:

> Hardly a week goes by without a prophecy of imminent disaster for the human race. If we are not to be poisoned by environmental pollution, we shall be starved through exhaustion of the soil, and if by some superhuman achievement of agricultural science food production looks like keeping pace with population growth, failure of water supplies will frustrate its success.

When I use the word "recently", in this instance I am speaking in geological terms, in which half a century is an eyeblink. The quote comes from an article in the *Financial Times* (Cherrington, 1972), shortly after the first

DOI: 10.4324/9781003461067-6

injection of carbon dioxide began in Texas, as described near the beginning of Chapter 3.

Concern about the consequences of industrialisation and the limits to the Earth's ability to either provide what we need or cope with "waste products" has waxed and waned since then. The prophecies of imminent disaster have returned, and alongside them have come claims of salvation by technology.

This book is being completed at a very interesting time, just before COP28 in Dubai. The International Energy Agency has released a report on the oil and gas industry and the energy transition (IEA, 2023). (The IEA has been releasing reports about CCS for a decade and a half; see IEA, 2009.) Interviewed about its launch, IEA's chief, Dr Fatih Birol, said that meeting the world's climate goals "means letting go of the illusion that implausibly large amounts of carbon capture are the solution" (Stallard, 2023). By the time you read this, it will be clear what – if any – impact such warnings have had on the negotiations and on national enthusiasms for more CCS.

The rest of this brief chapter proceeds as follows. In the next section, I very briefly summarise the empirical findings and point to common themes. In the following section, I offer some unsolicited advice to those studying and talking about CCS. In the final section, with great reluctance and caution, I offer some predictions about what may be coming.

6.2 Recap

Chapter 3 traced the very long pre-history of CCS, a three-decade process in which the idea was spoken of by a small band of academics and scientists. During this period, the "policy" and "politics" streams were either largely empty (pre-1988) or else concerned with other aspects of climate change, especially around responses that included denial and incremental and "economically sound" measures such as carbon pricing and carbon trading. Chapter 4, albeit largely focussed on the UK, showed that there had been repeated efforts by advocates of the technology to gain necessary state support and a long and repeated story of procrastination and delay by policymakers. They could agree in principle that "something should be done", but once decisions had to be made that would require sufficient precision to give investors the clarity required, policymakers would delay and prevaricate. Repeatedly, this has led to the collapse of proposals and competitions. This period culminated with a decision that was not, it seems, made with hostility to CCS but something that some probably regarded as more enraging – a kind of studied indifference. While the first half of Chapter 5 covers the remarkable comeback of CCS as an accepted mitigation option, the second half shows that the old habits, of deferring decisions about the fine details, persist. The policy window may have reopened, but the old problem of implementation has not gone away. There is now significant rhetorical support for CCS, and non-trivial amounts of state money have been dispensed. But until FIDs have been made (a topic returned to below), it would be a brave person who would say that CCS has been embedded in the UK and that it will survive future shocks and reversals.

6.3   Unsolicited advice for researchers and observers

In this section I offer a few bits of advice about how to gain information and some of the tools (or at the very least perspectives and heuristics) that you might use to analyse

6.3.1   Learn to sip from the fire hydrant

The problem for those with an interest in the CCS story is not scarcity but rather overabundance of information. Part of the way of dealing with this is to make an initial decision over what sources of information you are going to monitor closely and what your stance is towards those sources.

- If you have a Gmail account, then you can set up a Google Alert (e.g. "carbon capture and storage" and "United Kingdom"). You can choose the frequency with which you receive these alerts. One advantage of this is that you can also see what the sources are and ignore or explore accordingly.
- The CCSA has an excellent free weekly newsletter, which you can sign up for on their website. There are other organisations (Green Alliance, Aldersgate Group, Greenpeace, Oil Change International, etc.) also worthy of your attention.
- The coverage of energy issues – and CCS – in the *Financial Times* is, as you would expect, excellent. See also the *Energy & Climate Intelligence Unit*.
- There are specialist publications (check the references at the end of various chapters of this book) which offer detailed accounts of the ins and outs.
- The House of Commons research library and other parliamentary bodies (e.g. the Parliamentary Office of Science and Technology) do an excellent job of producing briefings (for MPs). These are an excellent leaping-off point.
- Google Scholar is a reasonable way of finding out what is being published by academics. You can then also search for where an article has been cited, in what time frames and what else the academic has produced.
- If academic papers are behind paywalls and you do not have institutional access, you could write a polite email to the authors – they may well be willing to let you have a copy of the article(s) in question.

It is crucial to take nothing as gospel. All authors and organisations (and this most definitely includes academics!) have their "angles", and their blindspots and prejudices; I tried to foreground mine at the end of Chapter 1. Some sources of information will have a very strong perspective – as the adage goes, "News is what someone wants suppressed. Everything else is just advertising." Over time you will develop a sense of who is able to tell you what, and how, and what things they are likely to overlook or occlude.

6.3.2   Capacity to act effectively – or "Brown M&Ms"

The US rock band Van Halen had a rider in their contract that if there were Brown M&Ms in the backstage hospitality room, they could refuse to perform the concert. While this was initially reported as an example of prima donna behaviour, there was in fact a compelling underlying logic. The concerts were vast enterprises, involving heavy equipment being moved frequently. The Brown M&Ms clause was there to provide a test of how effective and precise the project management protocols of the organisers were. If they were unable to get something relatively straightforward (albeit 'trivial') correct, why should they be trusted with much bigger tasks where the consequences of failure could in fact be fatal?

In 2023, if you were an investor of a nervous disposition, you would assess that the UK government has failed the Brown M&Ms test on two issues distinct from, but related to, CCS. First, it failed to set the parameters for the latest auction for offshore wind, which uses the much-loved "Contracts for Difference" mechanism (Crerar & Ambrose, 2023). This has already spooked investors (Smeeton, 2023). This is a new-ish but maturing technology and a favoured policy mechanism, but still, failure abounded.

The second case is still more extraordinary. The UK government had committed to the construction of a high-speed railway line between London and the North of England. Over the last few years, with enormous cost-overruns, its future came into doubt. Finally, in October 2023, it was cancelled. Without wanting to underplay the complexity of megaprojects, railways are a 19th-century technology with a very clear set of users and revenue streams. If the UK government cannot oversee this, it stretches charity to believe that any investor will have confidence in the construction of huge transport and storage infrastructure (and the accompanying regulations, legal frameworks, etc.) for CCS.

6.3.3   CCS as Rorschach test?

Some "biases" are obvious (it is relatively easy to see when someone is telling you a one-sided story in favour of or in opposition to CCS). Some will emphasise the upsides, others the downsides. The tendency towards magical thinking and extreme technophilia runs deep in Western society. Thinking of CCS as a kind of "Rorschach test" might be helpful. It is named after its inventor, a Swiss psychologist, Hermann Rorschach. He created various inkblot images and asked patients to see what they saw. The interpretations they gave were then analysed for insights into how the patient saw the world.

Those who see CCS as a vital component of delivering the carbon-free glass, steel, cement and other industrial products required to build a zero-carbon infrastructure will say that CCS has not been given sufficient support and stability to work. Those who see CCS as the latest false promise by the oil and gas industry, determined to keep pumping out oil and gas regardless of the consequences, will say that 20 years of failure is enough and that it is time to stop pretending. The author is still stuck in duck/hare mode.

6.3.4 *Ask questions*

Related to the above points, it is crucial to "interrogate" sources of information, be they articles, books or speakers at events. Some of my favourite questions are as follows (but it is a very incomplete list):

- If everything worked absolutely perfectly (and remembering it never does), how much would be the net gain from this technology? How does that relate to the scale of the problem we face?
- Have we been here before with this proposal or something very like it? If so, what happened last time? Is it possible or likely that something similar will happen this time?
- (Where) are we in the hype cycle? How is that affecting how this particular proposal is received? How likely is it that the specific project will bear fruit?
- Watch the Marge versus the Monorail episode of *The Simpsons* for insights into how people get swept up into exuberance about a technology which may in fact overpromise and underperform and the costs attached to pursuing it.

6.4 Predictions

The Danish Nobel Prize–winning physicist Niels Bohr once quipped that prediction was very difficult, especially about the future. By the time this book is published, many entirely predictable but only-in-retrospect things will have happened.

What is offered below should be read in that spirit.

6.4.1 *There will be fierce battles over funding*

At the time of writing (October 2023), it seems very likely that the Conservative Party will not form the next government, after a general election, which must be held by January 2025. While the Labour Party under Keir Starmer is making positive noises about CCS (Taylor, 2023; Carrell, 2023), with its "British Jobs Bonus", it seems to me that there will be a continued period of "hedging", with politicians unwilling or unable to make firm enough announcements to ensure that FIDs can be made by private companies before the election. After the election, if Labour wins, there will be a period of "settling in" and "reassessing". This could easily push back decisions about CCS in the UK into 2025–2026.

There will be close scrutiny, as well, over what is funded. In November 2023, the Institute for Energy Economics and Financial Analysis (IEEFA), an organisation that has long been critical of CCS, released a report (Reid & Flora, 2023) that looks at Track 1 of the UK government's £20 billion industrial decarbonisation programme, in which it has initially selected eight CCS projects in two clusters (HyNet and the East Coast Cluster). IEEFA argues that the government is disproportionately supporting the development of blue hydrogen made with fossil gas. It points out that money earmarked for industrial decarbonisation is actually being spent on a new gas-fired power plant. (For further details on the IEEFA report and the responses to it from the Department for Energy Security and Net Zero, see Keating, 2023.)

Given its wide but shallow support, CCS in the UK probably cannot afford even the *perception* of such misallocation of funding.

6.4.2 There will be battles over legitimacy and public acceptance of CCS

One of the main arguments that UK-based CCS advocates had was that there was a chance to "lead the world". There was a time when that may well have been true. However, over the last two or three years, that argument has been undercut by increased interest (and in some cases action) in the US, Europe and beyond. It may be, then, that the argument is retired and instead the idea of the UK as an importer and storage facility for other countries' carbon dioxide gains prominence. After all, CCS advocates in the United Kingdom have one advantage over those advocating in Europe and the United States – the transport and storage facilities are, for the most part, not near where anyone lives. This removes the immediate "not in my backyard" salience for opponents of such schemes.

These battles will involve re-branding of existing organisations, which will do their best to remove words like "oil", "gas" and "petroleum" from their names or, if that is not possible, add words like "sustainable", "renewable" or "green" to their brands. We have already seen this in 2018, when the Norwegian state-owned oil company changed from Statoil to the much more cuddly sounding Equinor. In March 2022, the Oil and Gas Authority became "The North Sea Transition Authority". In September 2023, the Australian Petroleum Production & Exploration Association became "Australian Energy Producers".

For some time, the social media accounts of various energy producers have been pushing out a constant stream of images of diverse (by gender, age and ethnicity) workers doing good deeds.

New uses and proposals will be found for sectors that would benefit from CCS, especially the "hard to abate" sectors. More specifically, one of the key tools of the CCS advocates will be the notion of utilisation. There are interesting parallels with EOR (see Chapter 3). Both hold out the promise of an answer to the questions: "Who is going to pay? Where is the *early* revenue going to come from?" In the early days (e.g. Miller Field, discussed in Chapter 4), EOR would have offered serious funding. However, EOR had little legitimacy (and surely less now), since ultimately more carbon dioxide would be vented into the atmosphere. Superficially, utilisation avoids this and would help with legitimisation/normalisation of CCS. However, given that a lot of the captured and used carbon dioxide would then be re-emitted in the *next* use, and is merely deferring rather than preventing eventual emissions into the atmosphere, its usefulness as a justification for CCS will be limited. To return to the bathtub analogy of Chapter 1, all that is being done is filling a cup, sitting it on the edge of the bathtub and then, a very short time later, tipping it back into the already almost-overflowing tub.

Some advocates of CCS will be put in an almost impossible situation in all this. They are aware of the dangers of overpromising and under-delivering, and also of the problem that CCS in general can be deployed as a fig leaf to support the ongoing exploration and extraction of oil and gas. However, picking that fight in public, challenging people and organisations that have provided funding and support in the past while trying to carve out a very specific and nuanced position in a media ecosystem that actively selects against nuance is unlikely to appeal to many such advocates. They will continue to do what they have done so far – bite their lip and hope that the clear case they have made for CCS for specific purposes (around hard to abate sectors) is not irreparably damaged by less scrupulous and well-intentioned speakers.

One final prediction – a quote from Occidental Petroleum CEO, Vicki Hollub, speaking at an Oil and Gas conference in March 2023 will be used repeatedly. She said:

> We believe that our direct capture technology is going to be the technology that helps to preserve our industry over time. This gives our industry a licence to continue to operate for the 60, 70, 80 years that I think is going to be very much needed.
>
> (Westervelt, 2023)

Five months later, Occidental acquired direct air capture pioneer Carbon Engineering for $1.1 billion (Dinova, 2023)

6.4.3 Accidents will happen

Academics and companies are pondering what constitutes public acceptability and a "social licence to operate" in the context of the resistance to the hydrogen village experiment in Whitby and the government decision to abandon it (Mavrokefalidis, 2023).

All this work about who will accept what remains relatively hypothetical, however.

All human endeavours involve accidents and unforeseen consequences. CCS, a vast enterprise, will be no different. There will be countless minor mishaps, some noticed, others not. If there is a serious accident involving the capture, transport or storage of carbon dioxide anywhere in the world, especially if it involves fatalities or the photogenic release of CO_2 from undersea storage, this will be seized on by opponents of the technology. They will seek to use it to delegitimise funding and approval for projects already underway or being planned. Their efforts, if successful, could easily reduce public support for CCS to such a point that governments and/or companies defer decisions or even cancel projects.

As John Underhill, academic executive director of the GeoNetZero Centre for Doctoral Training in Edinburgh, told journalists, "the geology is absolutely key too. If we pick the wrong ones and something goes awry then the industry will lose credibility and we may not get a second chance" (Smyth & Sheppard, 2021).

6.4.4 The US elections of 2024

Almost every US presidential election the author can remember has been billed as "crucial" or "pivotal". Some of them indeed were. The next is being billed as such, with perhaps more justification than most. The administration of the Democrat Joe Biden has created enormous incentives for CCS (and other technologies) in the UK. If former President Donald Trump regains the White House in November 2024, then it seems very likely that Biden's incentives – alongside US participation in international climate negotiations – will end (Luce, 2023; Smyth & Wiliams, 2023).

6.5 Conclusion

Readers with a good memory will remember that, as noted at the start of Chapter 1, the UK magazine *Private Eye* compared CCS to the Loch Ness Monster. The search for Nessie conducted in August 2023 revealed nothing, but the believers – and those who choose to believe for various reasons – are not deterred and will keep on generating words and activity around the fabled creature. True believers will not be deterred – absence of evidence is not, after all, evidence of absence. However, the search for a curious being which almost certainly does not exist is quite different from the search for solutions (technological, social, or "both") to the dilemmas facing our species.

There is a saying that *"all history repeats first time is tragedy, the second time as farce"*. Another says, *"history doesn't repeat, but she rhymes"*. A third, my personal favourite – *"every time history repeats, the price goes up"*. This book has argued for a historically informed understanding of what is at stake when we put several mitigation eggs in the CCS (or CCUS, or BECCS) basket. If action had been taken early, the price would have been manageable. The later action is left, the more expensive (and difficult) it gets. That action was not taken in the 1980s and 1990s has left us with complex, expensive and frankly quite speculative options on the table.

As Professor Kevin Anderson recently posted on Twitter (Anderson, 2023):

> Certainly responsibility is widespread. My concern with many senior climate academics is that we've made the political choice to sweeten the mitigation pill with so much tech (including highly speculative tech) that we've undermined the need to ask much deeper social questions.

For the reasons stated above, it seems unlikely that CCS will simply go away and that we will return to the pre-2003 days. It also seems unlikely that funding will be abruptly cut off as it was in 2007 and 2015. It is more likely, I suspect (and fear or hope, depending on whether I am seeing a duck or a hare), that CCS will kind of just fizzle out, and everyone will agree that it is not actually salvageable. If that happens, we will be left with inevitably increasing emissions and nothing technological to do about them. Globally, some form(s) of CCUS in some form will happen. It may be that the UK is able to sell storage but not be a "leader" in any other sense, lacking attractiveness to investors.

If that is the case, it will be another example of warnings unheeded, of a slow march towards failure. In 1967, a year to the day before he was assassinated, the US civil rights leader Martin Luther King gave a speech. It was his first speech about the Vietnam War, a topic he'd been avoiding until then. In that speech, he warned:

> We are confronted with the fierce urgency of now. In this unfolding conundrum of life and history, there is such a thing as being too late. . . . We may cry out desperately for time to pause in her passage, but time is adamant to every plea and rushes on. Over the bleached bones and jumbled residues of numerous civilizations are written the pathetic words, "Too late".
>
> (King, 1967)

References

Anderson, K. 2023. Tweet about academic complicity. *Twitter*, September 12. https://twitter.com/KevinClimate/status/1701560143287111820

Carrell, S. 2023. Starmer says £2.5bn renewables jobs fund will help North Sea oil workers. *The Guardian*, November 16. www.theguardian.com/

environment/2023/nov/16/starmer-plans-renewable-jobs-fund-north-sea-oil-workers

Cherrington, J. 1972. What the alarmists forget. *Financial Times*, April 4, p. 7.

Crerar, P. and Ambrose, J. 2023. Blow to Rishi Sunak as offshore wind auction appears to have zero bidders. *The Guardian*, September 7. www. theguardian.com/business/2023/sep/07/suank-offshore-wind-auction-bidders-government-energy-bills

Dinova, D. 2023. Occidental acquires Carbon Engineering for $1.1bn. *Carbon Herald*, August 26. https://carbonherald.com/occidental-acquires-carbon-engineering-for-1-1-billion/

IEA. 2009. *Technology Roadmap—Carbon Capture and Storage*. Paris: IEA. www. iea.org/reports/technology-roadmap-carbon-capture-and-storage-2009

IEA. 2023. *The Oil and Gas Industry in Energy Transitions*. Paris: IEA. www. iea.org/reports/the-oil-and-gas-industry-in-energy-transitions

Keating, C. 2023. Has the fossil fuel industry 'captured' the UK's CCS programme? *Business Green*, November 16. www.businessgreen.com/news-analysis/4146944/fossil-fuel-industry-captured-uks-ccs-programme

King, M. L. 1967. Beyond Vietnam. *Speech delivered at Manhattan's Riverside Church*, April 4. https://www2.hawaii.edu/~freeman/courses/phil100/17.%20MLK%20Beyond%20Vietnam.pdf

Luce, E. 2023. Reply to Pilita Clark's "The relentless rise of the IRA. *Financial Times*, July 10. www.ft.com/content/1b8a147d-88cd-43a0-9dd0-565c3d52b0fb

Mavrokefalidis, D. 2023. Hydrogen village trial in Whitby rejected. *Energy Live News*, July 11. www.energylivenews.com/2023/07/11/hydrogen-village-trial-in-whitby-rejected/

Reid, A. and Flora, A. 2023. UK carbon capture policy: out of step with net-zero goals. *Institute for Energy Economics and Fiscal Analysis*, November 15. https://ieefa.org/resources/uk-carbon-capture-policy-out-step-net-zero-goals

Smeeton, G. 2023. Billpayers could miss out on £1bn a year in savings due to wind auction error. *ECIU*, September 2023. https://eciu.net/media/press-releases/2023/billpayers-could-miss-out-on-1bn-a-year-in-savings-due-to-wind-auction-error

Smyth, J. and Sheppard, D. 2021. Monster problem: Gorgon project is a test case for carbon capture. *Financial Times*, July 26. www.ft.com/content/428e60ee-56cc-4e75-88d5-2b880a9b854a

Smyth, J. and Wiliams, A. 2023. Donald Trump would gut Joe Biden's landmark IRA climate law if elected. *Financial Times*, November 23. www.ft.com/content/ed4b352b-5c06-4f8d-9df7-1b1f9fecb269

Stallard, E. 2023. COP28 'moment of truth' for oil industry, says energy boss. *BBC*, November 23. www.bbc.co.uk/news/science-environment-67445233

Taylor, M. 2023. Starmer in Aberdeen to reveal further details of Labour energy plan. *Holyrood*, November 17. www.holyrood.com/news/view, starmer-in-aberdeen-to-reveal-further-details-of-labour-energy-plan

Westervelt, A. 2023. An "unabated" disaster. *Drilled*. https://drilled.ghost.io/an-unabated-disaster/?ref=drilled-newsletter

Index